PRACTICAL STATISTICS & PROBABILITY

PRACTICAL STATISTICS & PROBABILITY

ROBERT LOVEDAY

M.Sc. (*Sheffield*), F.I.S., F.I.M.A.

CAMBRIDGE
AT THE UNIVERSITY PRESS
1974

Published by the Syndics of the Cambridge University Press
Bentley House, 200 Euston Road, London NW1 2DB
American Branch: 32 East 57th Street, New York, N.Y.10022

ISBN: 0 521 20291 4

Printed and bound in Great Britain by
REDWOOD BURN LIMITED
Trowbridge & Esher

Contents

6. The Binomial Distribution

7. The Normal Distribution

8. Regression

9. Correlation by Product-moments

10. Correlation by Ranks

11. The Analysis of a Time-series

12. Weighted Averages

13. The Presentation and Collection of Data

14. Level of Significance

Acknowledgements

The author and publishers would like to thank the following Boards for permission to reproduce examination questions:

The University of Cambridge Local Examinations Syndicate (Cambridge.)

The Oxford and Cambridge Schools Examinations Board (O. & C.)

The University of London University Entrance and School Examinations Council (London.)

The Joint Matriculation Board, Manchester (J.M.B.)

The Associated Examining Board for the General Certificate of Education (A.E.B.)

The West African Examinations Council (W.A.E.C.)

The Mathematical Association

Preface

Nowadays, beginners in statistics generally wish to carry their studies beyond frequency distributions, location, dispersion, regression, correlation, time-series, weighted averages and the collection and representation of data. This book, therefore, covers all these topics but goes further. It introduces, in a simple way, sufficient probability to proceed to an understanding of the Binomial distribution and the way in which it tends to the Normal distribution. This leads, in the first place, to the construction and use of Normal probability graph paper and eventually to the discussion of level of significance. A capital letter is given to the word *Normal* whenever it is used with its special meaning and, to preserve uniformity, a capital letter is also given to *Binomial*. In an elementary introduction such as this, small samples, the divisor $(n-1)$, t, χ^2 and F tests have been omitted. These topics are fully covered in my *Statistics* (C.U.P.).

Awareness of the power of statistics is, today, part of general knowledge. Reports, discussions, advertisements are made convincing by presenting, in some suitable way, collections of facts and figures. For this reason precious hours can be wasted at the beginning of a course in general discussion. The time can be used to better advantage in the immediate introduction of the special techniques which are not common knowledge. In the first chapter of this course, then, the student is plunged into a new world of diagrams, terminology and calculations by means of a wide variety of examples. Any new word or phrase is introduced in *italics* to signify that it must be carefully studied and memorized. The presentation and collection of statistical data is finally used in Chapter 13 as a summary. By then the reader can cover the ideas quickly because he has already had considerable practice in the tabulation and graphical representation of observations and has already studied a table of random numbers.

In the first five chapters the student should gain enough experience and understanding to approach with confidence the difficult chapters 6, 7 and 8. However, it is possible to omit entirely chapters 5, 6, 7, 8, 9, 10 and 14 at the first reading because chapters 5 to 7, chapters 8 to 10 and chapter 14 form three separate independent sections.

The whole course carries the study of statistics as far as it is possible to go without a really good background in Pure Mathematics. It covers the requirements of

(i) G.C.E. Ordinary Level examinations in the separate subject Statistics;
(ii) statistics sections in many G.C.E. Advanced Level examinations;
(iii) statistics papers of O.N.D. and H.N.D. courses;
(iv) the first year of courses leading to a Degree in Business and Social Studies.

Answers are provided at the end of the book for all the exercises. The alphabetical Glossary not only defines the terms used in the text but gives brief introductions to more advanced ideas.

July 1973 R.L.

I

Frequency distributions

The method of this course is to teach the student by worked examples. Statistical data will be taken and analysed in order to illustrate certain well-known processes. Only, when the study of the example is complete, will the student realize what the example has taught.

1. The formation of a frequency distribution. Table 1 A gives the weights, in kg (to the nearest kg), of 32 men of one nationality and table 1 B gives the weights of 256 men of a different nationality. In the discussion which follows it will be convenient to refer to the two groups simply as *men A* and *men B*.

TABLE 1 A

Weights in kg (to the nearest kg) of 32 men A

57	62	58	58	56	59	51	56
60	56	62	57	58	61	57	61
60	60	58	57	60	55	56	62
57	71	64	61	53	63	54	67

First of all it should be noted that the weights have been *rounded-off* to the nearest kg. This means that the figure 57 is put down for the first man because his weight is between 56·500 kg and 57·499 kg *inclusive* and the figure 58 is put down for the third man because his weight is between 57·500 kg and 58·499 kg *inclusive.* This assumes that the weighing machine is capable of registering 0·001 kg. If it is only capable of registering 0·1 kg the *rounded* figure of 57 would indicate between 56·5 kg and 57·4 kg *inclusive.* In other words 57 indicates

greater than or equal to $56\frac{1}{2}$ *but less than* $57\frac{1}{2}$.

In mathematical symbols we write this as

$$56\tfrac{1}{2} \leqslant X < 57\tfrac{1}{2}$$

where X is the weight of the first man.

TABLE 1B

Weights in kg (to the nearest kg) of 256 men B

66	67	75	76	67	76	76	79	78	84	79	79
77	72	70	70	73	78	81	71	81	83	76	76
71	75	81	67	73	74	73	79	71	82	76	70
83	72	85	70	78	76	80	77	89	80	71	78
75	73	72	88	66	69	73	77	86	86	77	79
85	72	75	84	66	77	75	79	90	83	81	77
72	74	75	81	70	65	81	88	76	85	65	70
81	78	80	74	74	75	76	78	84	78	77	84
64	75	70	84	73	66	77	79	75	73	78	73
87	81	76	71	67	73	73	70	79	85	69	68
69	70	86	80	83	81	79	80	95	86	89	80
80	86	85	86	79	75	81	77	95	75	70	66
81	67	77	69	68	78	80	83	81	80	80	78
86	77	83	71	67	76	80	81	83	85	77	72
73	68	73	81	73	75	79	83	85	80	80	92
79	80	73	74	65	77	79	78	82	76	76	66
72	74	72	76	75	81	78	73	71	82	92	95
77	86	69	70	74	71	80	75	89	80	70	67
68	64	71	84	100	77	64	78	70	73	74	79
72	81	73	66	68	62	68	84	68	102	75	69
70	72	73	81	78	75	79	67	69	71	79	76
72	74	77	62								

TABLE 1C

The method of forming a frequency distribution from the weights of men A given in table 1A

	Column 1	Column 2	Column 3	Column 4
	Weights grouped in 5 kg intervals (nearest kg)	Weight intervals kg		Frequency
Row 1	50, 51, 52, 53, 54	$49\frac{1}{2}$–$54\frac{1}{2}$	III	3
Row 2	55, 56, 57, 58, 59	$54\frac{1}{2}$–$59\frac{1}{2}$	~~IIII~~ ~~IIII~~ ~~IIII~~	15
Row 3	60, 61, 62, 63, 64	$59\frac{1}{2}$–$64\frac{1}{2}$	~~IIII~~ ~~IIII~~ II	12
Row 4	65, 66, 67, 68, 69	$64\frac{1}{2}$–$69\frac{1}{2}$	I	1
Row 5	70, 71, 72, 73, 74	$69\frac{1}{2}$–$74\frac{1}{2}$	I	1
			Total	32

Table 1 C indicates the process of summarizing the data of table 1 A. The 32 weights have been divided into groups of 5 kg intervals, *row* 1 showing the number of men whose weights are 50, 51, 52, 53 or 54 kg (to the nearest kg), that is, between $49\frac{1}{2}$ and $54\frac{1}{2}$ kg; *row* 2 showing the number of men whose weights are 55, 56, 57, 58 or 59 kg (to the nearest kg), that is, between $54\frac{1}{2}$ and $59\frac{1}{2}$ kg and so on. To build up the table we take the weights 57, 62, 58, ... one by one and place a small

TABLE 1D

Frequency distribution of the weights of men B

Weight (kg)	Frequency
$59\frac{1}{2}$–$64\frac{1}{2}$	5
$64\frac{1}{2}$–$69\frac{1}{2}$	32
$69\frac{1}{2}$–$74\frac{1}{2}$	61
$74\frac{1}{2}$–$79\frac{1}{2}$	78
$79\frac{1}{2}$–$84\frac{1}{2}$	51
$84\frac{1}{2}$–$89\frac{1}{2}$	21
$89\frac{1}{2}$–$94\frac{1}{2}$	3
$94\frac{1}{2}$–$99\frac{1}{2}$	3
$99\frac{1}{2}$–$104\frac{1}{2}$	2
Total	256

stroke for each of them in the appropriate *row* of *column* 3. Every fifth stroke is drawn diagonally across the preceding four strokes to form small groups of five. This enables us to see at a glance the total for each row when we enter it in *column* 4. We see, therefore, that *column* 4 gives the *frequency* with which the weights in each group occur. Table 1 C is an example of a *frequency distribution.*

The student should now go through the process of table 1 C using the weights of the 256 men *B* given in table 1 B. The correct frequencies giving a total of 256 are shown in table 1 D.

An alternative way of setting out table 1 C is shown in table 1 E. Columns 1 and 3 have been discarded because they are only necessary for describing and carrying out the process of forming the table. The method of writing the *group intervals* $49\frac{1}{2}$–, $54\frac{1}{2}$– and so on stresses the fact that the weights of the 3 lightest men are greater than or equal to $49\frac{1}{2}$ but less than $54\frac{1}{2}$, the weights of the next 15 men are greater than or equal to $54\frac{1}{2}$ but less than $59\frac{1}{2}$ and so on.

TABLE 1E

Alternative version of the frequency distribution of table 1 C

Weight in kg	$49\frac{1}{2}$–	$54\frac{1}{2}$–	$59\frac{1}{2}$–	$64\frac{1}{2}$–	$69\frac{1}{2}$–	Total
Frequency	3	15	12	1	1	32

2. The histogram. It is customary to represent a frequency distribution diagrammatically in the form of a *histogram* as shown in fig. 1. Here two separate histograms have been drawn to compare the distribution of the weights of men *A* with the distribution of the weights of men *B*. It is important to note that:

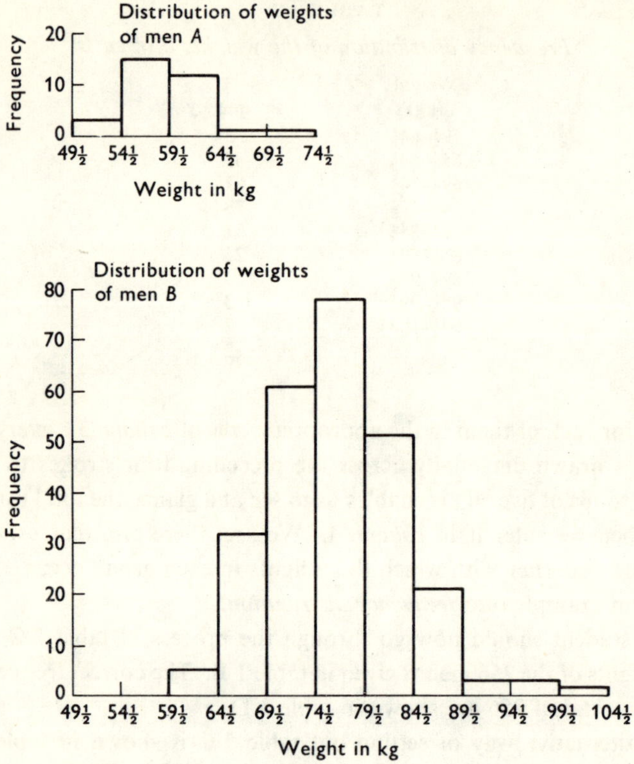

Fig. 1. Two separate histograms placed in such a way that the weights of men *A* can easily be compared with the weights of men *B*.

(i) To make a true comparison the scales of the axes of the two histograms are the same and one origin is directly above the other.

(ii) The separate frequencies are represented by rectangles whose areas are proportional to the frequencies.

(iii) The total areas of the two histograms are proportional to the totals 32 and 256 of the two groups of men.

(iv) Although the histograms do not tell us more than the frequency distributions, they help to complete our understanding.

(v) The histograms show diagrammatically that the weights of men *A range* from 49½ to 74½ kg with most weights between 54½ and 64½ kg while the weights of men *B range* from 59½ to 104½ kg with most weights between 69½ and 84½ kg. In other words, *A* is a nation of small men and *B* a nation of big men.

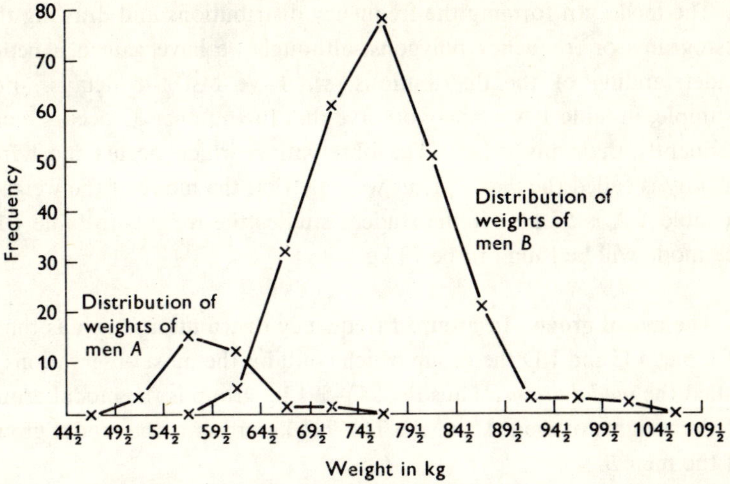

Fig. 2. Two frequency polygons drawn in the one diagram so as to compare the weights of men *A* with the weights of men *B*.

3. The frequency polygon. In fig. 2 two *frequency polygons* have been drawn in the one diagram to compare the two distributions. The scales of the axes have been kept the same as fig. 1 so that the student can compare the frequency polygons with the histograms. Note that

(i) The frequency polygon is formed by the straight lines joining points plotted above the mid-points of the group intervals.

(ii) The frequency polygon is completed at each end plotting zero points at 47 and 107.

(iii) The shapes of the frequency polygons portray the nature of the two distributions just as clearly as the histograms and they are more economical because they do it in one diagram.

(iv) Because the scales of fig. 2 are the same as fig. 1, the total area of each frequency polygon is equal to the total area of each histogram.

(v) The frequency polygons suggest that the weights of men *A* range from 47 to 77 kg while the weights of men *B* range from 57 to 107 kg. This must not be regarded as a contradiction of the inference made in § 2, (v), from the histograms. It is really a prediction. If many more men of each race were weighed, a few, but very few, might be found to be heavier and lighter than those of the 32 and 256.

(vi) It is not necessary to draw both the histogram and the frequency polygon to represent a distribution. One or the other is sufficient.

4. The mode. In forming the frequency distributions and drawing the histograms or frequency polygons, although we have gained a better understanding of the distributions, we have lost the details. For example, in table 1 A we may observe that the number 57 occurs more frequently than any other. The observation which occurs most frequently is called the *mode*. Thus we state that the mode of the weights in table 1 A is 57 kg. If the student studies the weights in table 1 B, the mode will be found to be 73 kg.

5. The modal group. In grouped frequency distributions such as those of tables 1 C and 1 D the group which contains the most observations is called the *modal group*. Thus the $54\frac{1}{2}$–$59\frac{1}{2}$ kg group is the modal group of the weights of men *A* and the $74\frac{1}{2}$–$79\frac{1}{2}$ kg group is the modal group of the men *B*.

The student may be surprised that the mode 73 kg found from table 1 B falls outside the modal group $74\frac{1}{2}$–$79\frac{1}{2}$. It should be realized, however, that the 256 weights of table 1 B are a *sample* of the *total population* of weights of all men of nation *B* and the histogram or frequency polygon gives a picture of what may be expected of the whole population. So that although 73 kg happens to be the mode of the sample we may expect the mode to be between $74\frac{1}{2}$ and $79\frac{1}{2}$ kg if all the thousands of men who constitute the nation *B* are weighed individually.

The use of a sample to estimate a measure for the whole population is, perhaps, the most important use of the study of statistics. The estimation of the mode will be discussed further in § 12 and § 38.

6. Experiments with dice. The histograms or frequency polygons obtained from the measurements of natural phenomena generally approximate in shape to certain well-defined curves called *frequency curves*. It is an amusing and useful exercise for the beginner to establish the shapes of these curves by throwing dice.

7. Throwing three dice: Normal distributions. It can be shown mathematically that, if three dice are cast, the probability of getting three sixes is 1/216. This means that there is 1 chance in 216 of scoring a total of 18. Further, the probability of getting two sixes and one five is 3/216. This means that there are 3 chances in 216 of scoring a total of 17. The following table is a complete list of the chances for the various possible total scores:

Total score	3	4	5	6	7	8	9	10
Chances in 216 of getting the score	1	3	6	10	15	21	25	27
Total score	11	12	13	14	15	16	17	18
Chances in 216 of getting the score	27	25	21	15	10	6	3	1

Fig. 3 shows the frequency curve obtained by drawing a graph from the above table. If the student actually throws three dice a large number of times and records the scores as shown in table 1 C he will obtain a frequency distribution which will give a histogram or frequency polygon approximating in shape to the *Normal frequency curve.* The author supervised the throwing of three dice 4320 (i.e. 20×216) times by twenty boys and obtained the following distribution, the frequency polygon of which approximates very closely in shape to fig. 3:

Score	3	4	5	6	7	8	9	10
Frequency	19	52	93	194	285	411	473	510
Score	11	12	13	14	15	16	17	18
Frequency	541	520	438	333	218	149	68	16

8. Throwing six dice: skewed distributions. If six dice are cast, it can be shown by the binomial theorem that the probability of getting no sixes is approximately 72/216, while that of getting one six is approximately 87/216. These and other values have been used in the drawing of fig. 4 which shows the shape of curve associated with a *skewed* distribution. When the curve has a long tail on the right of the mode and a short tail on the left it is said to be skewed positively. The student will also meet cases of negative skewness in which the long tail is on the

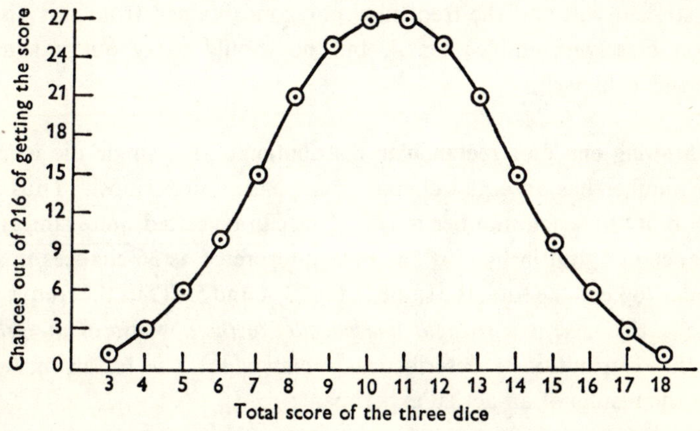

Fig. 3. Approximately Normal distribution. (A truly Normal curve is rather more sharply peaked rising to 29 instead of 27).

left of the mode. *Skewness* is, therefore, a term which distinguishes distributions from *Normal* distributions. The latter are symmetrical about the mode but if a distribution is symmetrical it is not necessarily Normal. The curve in fig. 3 is symmetrical but it is only an approximation to the truly Normal curve whose mathematical equation is given in § 91. The rectangular distribution shown in fig. 5 is another example of a symmetrical distribution which is not Normal.

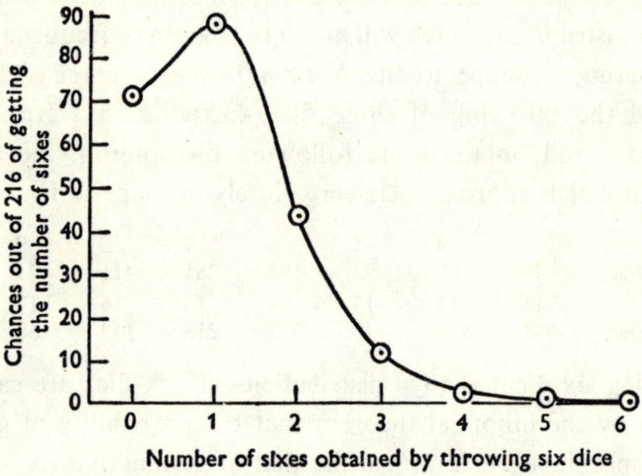

Fig. 4. Skewed distribution (skewed positively).

The following distribution was obtained in an actual experiment:

No. of sixes	0	1	2	3	4	5	6
Frequency	77	84	34	19	2	0	0

The student will find the frequency polygon obtained from this distribution closely resembles fig. 4, but he should carry out a similar experiment himself.

9. Throwing one die: rectangular distributions. If a single die is cast, each number has an equal chance of appearing uppermost. Thus the probability for each number is 1/6. This can be stated quite simply as 1 chance in 6, but in fig. 5 it has been interpreted as 36 chances in 216 in order to keep the total the same in figs. 3, 4 and 5. Thus the frequency curve in this case is a *straight line parallel to the abscissa of the graph* and the distribution is described as *rectangular*. The following table gives the results of an actual experiment:

Score	1	2	3	4	5	6
Frequency	96	98	117	130	107	100

The student should carry out a similar experiment for himself. In these experiments it is assumed, of course, that the dice are *true* (i.e. that they are not *loaded* or *biased*).

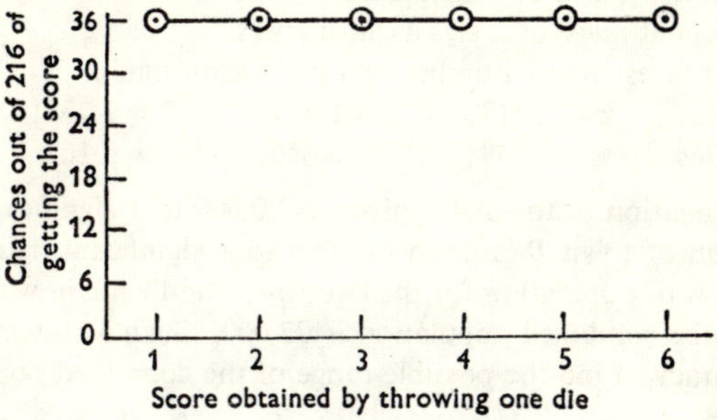

Fig. 5. Rectangular distribution.

10. Relative frequency. The ideas of §§ 6–9 lead to the writing of tables 1 C and 1 E in the form shown in table 1 F. In it, each frequency has been divided by the total number of observations, 32, to give the *relative frequency*. The total of the relative frequencies is, therefore, *unity*.

TABLE 1F

Relative frequency distribution of the weights of men A

Weight in kg	49½–54½	54½–59½	59½–64½	64½–69½	69½–74½	Total
Relative frequency	$\frac{3}{32}$	$\frac{15}{32}$	$\frac{3}{8}$	$\frac{1}{32}$	$\frac{1}{32}$	1

From table 1 F, the student will realize that for the whole nation *A* it is likely that $\frac{3}{8}$ of the men weigh between 59½ and 64½ kg, and similarly $\frac{3}{32}$ of the men weigh between 49½ and 54½ kg. Thus the relative frequency distribution takes our mind away from the original sample of observations and provides for us estimated fractions of the whole population. It is an easy step to multiply the relative frequencies by 100 and obtain the percentage frequency distribution of table 1 G.

TABLE 1G

Percentage frequency distribution of the weights of men A

Weight in kg	49½–54½	54½–59½	59½–64½	64½–69½	19½–74½	Total
Percentage frequency	9	47	38	3	3	100

11. Exercises.

1. The following numbers have been taken from a book of 4-figure tables. They are correct to 3 decimal places or 4 significant figures. Round them off
   (i) to 2 decimal places or 3 significant figures,
   (ii) to 1 decimal place or 2 significant figures,
   (iii) to the nearest whole number or 1 significant figure.

$$\pi = 3{\cdot}142; \quad e = 2{\cdot}718; \quad \sqrt{2} = 1{\cdot}414; \quad \sqrt{3} = 1{\cdot}732;$$
$$\sqrt{5} = 2{\cdot}236; \quad \sqrt{6} = 2{\cdot}449; \quad \sqrt{7} = 2{\cdot}646; \quad \sqrt{10} = 3{\cdot}162.$$

2. The population of town A is given as 20,000 to 1 significant figure and the population of town B is given as 2000 to 1 significant figure. State the possible ranges of population for the two towns and explain why it is wrong to state that the combined population is 22,000. Such a statement is called *spurious* accuracy. Find the possible range of the combined population.

3. The following table gives the heights, in cm (to the nearest cm) of 100 men. By using group intervals

154·5–159·5 cm, 159·5–164·5 cm, ..., 194·5–199·5 cm

form a frequency distribution and draw the histogram. Find the mode of the 100 observations and state the modal group of the distribution.

170	165	185	179	175	167	171	183	173	183
165	177	186	183	172	176	157	177	197	171
185	186	176	183	160	174	176	196	178	168
174	180	183	173	177	162	163	172	169	194
180	178	170	186	166	178	177	164	193	171
177	187	180	176	182	172	189	176	177	168
166	172	187	180	172	182	179	182	182	173
181	184	177	184	184	177	182	177	168	193
175	181	181	181	188	188	167	189	192	179
184	172	177	188	179	170	171	175	190	191

4. The following is a record of marks obtained by a group of students in an examination:

25	63	82	71	12	57	63	38	17	23	0	96	90	62
54	19	70	45	70	44	43	18	93	2	15	60	20	52
25	42	40	70	63	62	83	18	27	58	50	52	47	44
23	81	70	58	31	32	28	19	23	72	58	37	76	64
48	63	62	58	38	35	37	46	81	73	75	63	61	53
81	71	53	33	65	72	82	89	30	53	35	3	50	71

By grouping the marks in the intervals 0–9, 10–19, ..., 90–99, form a frequency distribution and draw the histogram. State the modal class.

By examining the original marks determine the mode.

5. Thirty students took two papers in an examination and their percentage marks were as follows:

Paper I	59	59	34	54	56	43	72	40	51	48
Paper II	56	74	36	54	81	39	67	62	42	62
Paper I	53	52	51	59	48	58	48	64	42	57
Paper II	43	83	77	55	71	49	42	58	42	58
Paper I	65	58	34	45	65	41	39	46	57	57
Paper II	52	63	35	48	54	43	29	43	54	73

Using class intervals 10–19, 20–29, 30–39, ... form two separate frequency distributions for the marks of Paper I and Paper II respectively and superpose in one diagram the two corresponding frequency polygons.

(i) Obtain the range of marks for each paper.

(ii) If the *pass* mark was 40, state the number of students who passed in each paper.

(iii) If the letter *A* was awarded for a mark of 70 or more, state, for each paper, how many students obtained *A*.

6. The following are populations of 100 village communities in a certain country area:

20	155	138	42	102	112	17	157	188	23
111	74	153	140	4	115	150	67	154	30
110	186	94	148	178	49	154	158	30	110
31	160	118	22	137	136	78	33	238	40
591	15	66	665	93	292	42	12	889	244
577	599	18	617	719	17	294	248	28	257
84	470	765	23	338	325	49	52	404	49
340	30	393	357	38	345	344	70	316	206
396	450	59	491	423	69	936	432	15	212
32	253	237	10	264	249	27	252	212	29

By grouping the populations in the intervals 0–99, 100–199, ... form a frequency distribution and construct the histogram. Note that it is shaped rather like a *reversed letter J*. State which is the modal group.

7. The following table gives the number of deaths occurring in each age group during 1972 for a certain community (deaths of persons of 70 or over have been omitted):

Age group	0–9	10–19	20–29	30–39	40–49	50–59	60–69
Deaths	23	9	3	1	1	11	42

Construct the histogram and note its shape. It is an example of a *U-type distribution.*

8. When 100 new bicycle frames were examined for flaws such as rough patches in the metal and scratches in the paintwork it was found that 58 frames had no flaws, 29 had 1, 8 had 2, 2 had 3, 2 had 4 and 1 frame actually had 5 flaws. Represent this statement as a histogram and note that it is an example of the *reversed-J type.*

9. *Weights of 100 dogs*

Weight in kg	0–5	5–10	10–15	15–20	20–25	25–30	30–35	35–40	40–45	45–50	50–55
Number of dogs	9	22	19	15	11	10	6	4	2	1	1

Illustrate the above frequency distribution by a frequency polygon. Note that the long tail to the right of the modal group indicates that the distribution is *skewed positively.*

10. Motor vehicles going into a certain workshop for repairs may take from 1 to 10 days before they are ready for use again. The following frequency distribution summarizes the times during which 100 vehicles were undergoing repairs:

Number of days taken to repair vehicle	1	2	3	4	5	6	7	8	9	10
Number of vehicles	25	17	10	7	6	1	2	5	10	17

Draw the histogram and explain how its shape may be due to the following factors:

(i) major and minor repairs;

(ii) time taken to discover faults;

(iii) time waiting for the arrival of spare parts;

(iv) time waiting for paint to dry on repairs to body work;

(v) time waiting for the owner of the vehicle to agree to the repairer's estimate of the cost.

11. Two smooth round stones were selected for this experiment, one weighing 150 g and the other weighing 4 kg. A hundred students were allowed to throw each stone 3 times and the best of their three distances for each stone were recorded. The following frequency distributions summarize the results:

Stone weighing 150 g

Distance in m	45–50	50–55	55–60	60–65	65–70	70–75	75–80	80–85	85–90	90–95
Number of Students	0	1	2	10	30	34	16	5	2	0

Stone weighing 4 kg

Distance in m	5–7½	7½–10	10–12½	12½–15	15–17½	17½–20	20–22½	22½–25
Number of Students	0	3	13	33	34	12	5	0

Illustrate the data by drawing two frequency polygons in one diagram and state the range and modal group of each distribution.

12. The following is a remarkably easy and quick experiment to carry out provided a stop-watch is available. Students throw a smooth round stone as high in the air as they possibly can. The time, in seconds, from the stone leaving the student's hand to its reaching the ground is recorded. The horizontal distance travelled by the stone is disregarded. The result of such an experiment when 100 students threw a 150 g stone is summarized by the following frequency distribution:

Time of flight of stone in seconds	2·0–2·5	2·5–3·0	3·0–3·5	3·5–4·0	4·0–4·5	4·5–5·0	5·0–5·5
Number of Students	0	4	20	38	28	9	1

Illustrate by a frequency polygon.

Note that, by the interval 3·5–4·0 in the above table, we mean a time of

3·5 sec or more but less than 4·0 sec. Using mathematical symbols this would be written

$$3{\cdot}5 \text{ sec} \leqslant \text{time of flight} < 4{\cdot}0 \text{ sec},$$

similarly for the other group intervals.

13. Turn to chapter 14 and examine the 400 random sampling numbers (excluding the means) given in table 14 B. From them form a frequency distribution with groups

$$0, 1, 2, 3, \ldots, 8, 9$$

and draw the histogram. It is an example of a rectangular distribution. Note that, even though the probability of each number is stated as $\frac{1}{10}$, the frequencies vary above and below 40 in the same way that, in § 9, the frequencies varied above and below 108.

14. Use the means in table 14 B to form a frequency distribution with the following groups:

$$2{\cdot}0\text{–}2{\cdot}4, \quad 2{\cdot}5\text{–}2{\cdot}9, \quad 3{\cdot}0\text{–}3{\cdot}4, \quad \ldots, \quad 6{\cdot}5\text{–}6{\cdot}9.$$

Draw the histogram and note that, unlike Ex. 13 above, the distribution is not rectangular.

12. Estimation of the mode. Although the modal group is easily obtained from the frequency distribution or histogram the mode, itself, is not so obvious. The following example illustrates two ways of *estimating* it:

(i) by a simple geometrical construction in the histogram;

(ii) by calculation from the frequency distribution.

The following table shows the frequency distribution of the marks of 800 candidates in an examination:

Marks	1–10	11–20	21–30	31–40	41–50
No. of candidates	10	40	80	140	170
Marks	51–60	61–70	71–80	81–90	91–100
No. of candidates	130	100	70	40	20

The histogram is shown in fig. 6. Note that the bases of the rectangles extend from $\frac{1}{2}$ to $10\frac{1}{2}$, $10\frac{1}{2}$ to $20\frac{1}{2}$, etc., so as to *enclose* the ranges 1–10, 11–20, etc. The modal group is 41–50. It is represented by the rectangle with base extending from $40\frac{1}{2}$ to $50\frac{1}{2}$. It contains 30 more candidates than the class below it and 40 more than the group above it. We therefore argue that the mode is *likely* to divide the modal group into the ratio 30:40. Hence we *estimate* the mode as

$$40\tfrac{1}{2} + \tfrac{3}{7} \text{ of } 10 \quad \text{or} \quad 50\tfrac{1}{2} - \tfrac{4}{7} \text{ of } 10,$$

and we write

$$\text{mode} = 44{\cdot}8,$$

which indicates that, if the marks are integers, the mode is likely to be 44 or 45.

Fig. 6 shows a geometrical method of determining the mode using the properties of similar triangles.

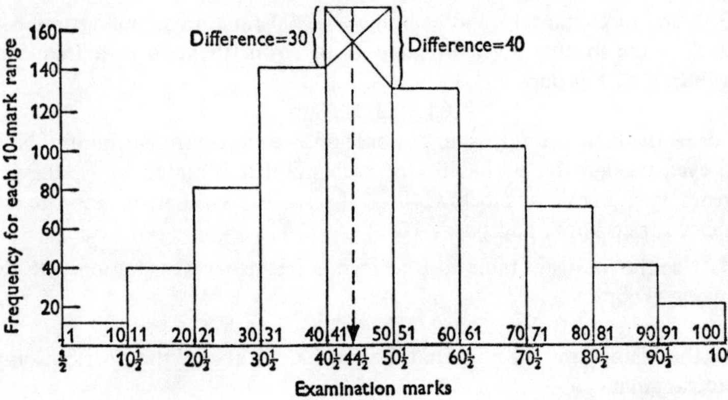

Fig. 6

13. Exercises.

1. The following is the frequency distribution of the weights, to the nearest kg, of 100 men:

Wt. (kg)	Frequency	Wt. (kg)	Frequency
50–54	1	80–84	17
55–59	2	85–89	10
60–64	5	90–94	6
65–69	11	95–99	4
70–74	21	100–104	2
75–79	20	105–109	1

Construct the histogram and find the mode. (London.)

2. In estimating the value of a plantation of trees, the girths of the trees in a sample area of 500 trees were measured and tabulated with a 0·2 m grouping interval as follows:

Girth (m)	0·4–0·6	0·6–0·8	0·8–1·0	1·0–1·2	1·2–1·4	1·4–1·6	1·6–1.8
No. of trees	25	30	135	160	100	40	10

Construct a histogram and estimate the mode. (London.)

14. Frequency distribution with unequal group intervals. Table 1 H is the percentage age distribution of the male population of a country 3 years after the end of a war. Note particularly the unequal group

TABLE 1H

Percentage age distribution of the male population of a country after the end of a war

Age group	Percentage of total population
Under 1	1·8
1 but under 2	1·7
2–4	4·8
5–9	7·5
10–14	8·0
15–19	8·5
20–39	24·0
40–59	26·0
60–99	17·7
Total	100·0

TABLE 1J

Bases and heights of the rectangles of the histogram representing the percentage age distribution of table 1H

Age group	Area of rectangle (percentage)	Base	Height
Under 1	1·8	1	1·8
1 but under 2	1·7	1	1·7
2–4	4·8	3	1·6
5–9	7·5	5	1·5
10–14	8·0	5	1·6
15–19	8·5	5	1·7
20–39	24·0	20	1·2
40–59	26·0	20	1·3
60–99	17·7	40	0·4425

intervals because they must be taken into account when the histogram is constructed. The first group 'under 1' represents an interval of 1 year, the second group '1 and under 2' also represents an interval of 1 year, but '2–4' is 3 years, '5–9', '10–14', '15–19' are each 5 years '20–39', '40–59' are each 20 years and '60–99' is 40 years. Table 1 J indicates how the rectangles of the histogram must be constructed so that their *areas* represent the *percentages*. The base of each rectangle represents the time interval of each group and the height of each rectangle is

$$\frac{\text{the percentage}}{\text{the time interval}}.$$

The histogram is shown in fig. 7.

15. Comments on the special features of the histogram. The histogram shown in fig. 7 has the following special features:

(i) The population of the '1 but under 2' group is less than that of the 'under 1' group. This is mainly due to the fact that it is usual for a number of infants to die before they reach the age of 1 year. The number of *deaths under 1 per thousand live births* is called the *infant mortality rate.*

(ii) The population of the '5–10' group is very low compared with the remainder of the child population. This group was actually '2–7' at the end of the war. It is probably low due partly to a high infant mortality and partly to a low birth rate during the war.

(iii) The '20–40' group is low. Normally we should expect that, due to deaths from natural causes there would be fewer men in the '40–60' group than in the '20–40'. The latter group, however, is that of the men who fought in the war and it portrays the loss of life.

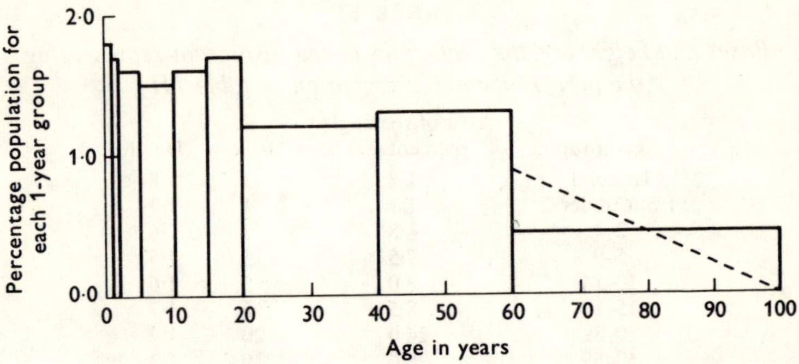

Fig. 7. Age distribution of the male population of a country 3 years after a war.

(iv) The '60–100' group is the low population due to old age. Although the rectangle represents the total population in this group it tends to make one think that there are as many men between 90 and 100 as there are between 60 and 70. The dotted line representing the last group by a triangle instead of a rectangle gives a better impression of the decrease in population due to old age.

16. Exercises.

1. The following table gives the age distribution of the population of a country:

Age	0–4	5–14	15–19	20–34	35–49	50–64	65–74	75–84	Total
Percentage of total population	10·0	19·0	8·3	19·8	18·6	15·0	6·1	3·2	100

Illustrate the data by a histogram.

2. The following table gives the distribution by seating capacity of the buses and coaches in a certain country:

Seating capacity	15–20	21–26	27–32	33–40	41–48	49–56	Total
Percentage of buses and coaches	3	2	4	20	69	2	100

Construct a histogram to illustrate the distribution.

3. *Frequency distribution of the diameters of the craters on the surface of the moon*

Diameter of crater (km)	Number of such craters (frequency)
5·0– 9·9	650
10·0– 19·9	438
20·0– 29·9	223
30·0– 49·9	163
50·0–149·9	122
Total	1596

Illustrate the above data by a histogram.

4. The following table gives the number of deaths occurring in each age group during 1952 for a certain village of population 6200:

Age group	0–5	5–15	15–25	25–35	35–45	45–55	55–65	65–85
Deaths	13	1	2	3	3	8	10	22

Construct a histogram. (London.)
Note the shape of the histogram. It is an example of a *U-type distribution.*

5. *Total population (male and female) of a country 20 years after a 6-year war and 35 years after a trade depression*

Age group	Population (thousands)	Age group	Population (thousands)
Under 1	941	40–44	3603
1 and under 2	943	45–49	3365
2–4	2767	50–54	3529
5–9	4166	55–59	3445
10–14	3796	60–64	3056
15–19	4226	65–69	2438
20–24	3574	70–74	1806
25–29	3241	75–79	1249
30–34	3186	80–84	713
35–39	3354	85 and over	388

Represent the above age distribution as a histogram and comment on its special features. ('Assume that 85 and over' means 85–100.)

2

Cumulative frequency distributions

17. The cumulative frequency distribution. The frequency distribution of the weights of men *B* given in table 1 D can be converted to a *cumulative frequency distribution* by adding each frequency to the total of its predecessors. An appropriate way of tabulating it is given in table 2 A.

TABLE 2A

Cumulative frequency distribution of the weights of men B

Weight in kg	Cumulative frequency
Less than 59½	0
Less than 64½	5
Less than 69½	37
Less than 74½	98
Less than 79½	176
Less than 84½	227
Less than 89½	248
Less than 94½	251
Less than 99½	254
Less than 104½	256

18. The cumulative frequency polygon or ogive. The graph of the cumulative frequency distribution shown in fig. 8 is called the *cumulative frequency polygon* or *ogive*. The word 'ogive' is a term used in architecture to describe curves shaped like that in fig. 8.

The shape of the polygon in fig. 8 indicates that the distribution of the weights of men *B* is approximately *Normal*. Distributions which are not Normal produce ogives quite different in shape from that of fig. 8.

The frequency polygon is an extremely useful graph. Many inferences can be drawn from it. The reader should make an accurate copy of fig. 8 on graph paper so that it can be used to verify the following statements:

(i) The ogive shows that 128 men weigh less than 76·4 kg; in other words, half of the men weigh less than 76·4 kg and half of them weigh more than 76·4 kg. This weight, 76·4 kg, is called the *median* of the distribution. It is also called the *50th percentile* because 50% of the men weigh less than 76·4 kg.

(ii) The ogive shows that 64 men weigh less than 71·7 kg; in other words one-quarter or 25% of the men weigh less than 71·7 kg. This weight, 71·7 kg, is called the *lower quartile* or *25th percentile* of the distribution.

(iii) The ogive shows that 192 men weigh less than 81·1 kg. in other words three-quarters or 75% of the men weigh less than 81·1 kg. This weight, 81·1 kg, is called the *upper quartile* or *75th percentile* of the distribution.

(iv) From (ii) and (iii) above, half the weights are between 71·7 kg and 81·1 kg. This range, 71·7–81·1 kg, is called the *interquartile range* or the *central 50%* of the weights.

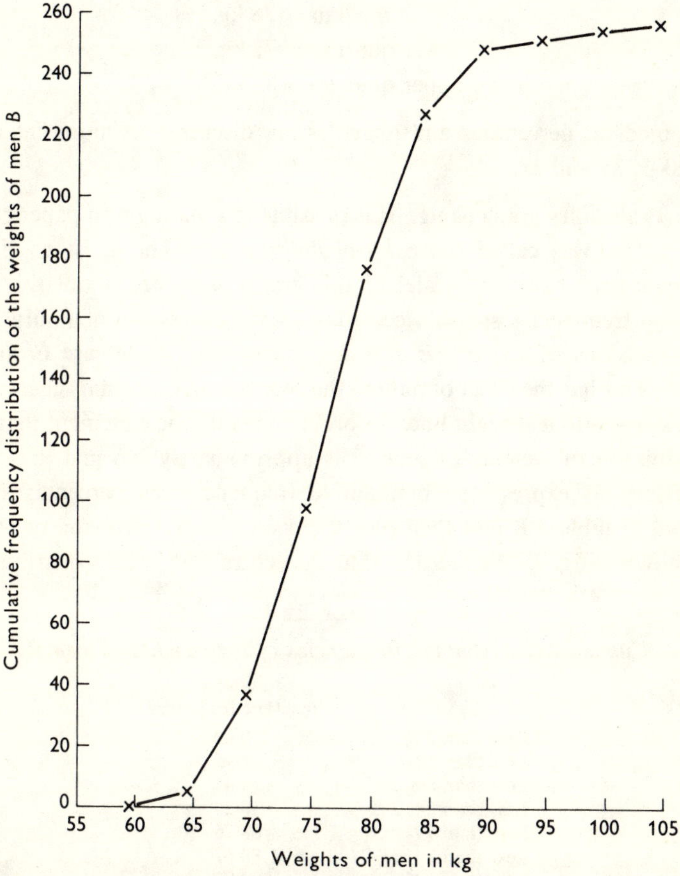

Fig. 8. The cumulative frequency polygon or ogive.

The median could have been deduced arithmetically from the middle rectangle of the histogram in fig. 1 by cutting off 30 from the 78 weights it contains in order to make the sum, 98, of the first three rectangles into 128. Thus the median is

$$74\tfrac{1}{2} + \tfrac{30}{78} \text{ of } 5 = 76{\cdot}4.$$

In the same way the quartiles can be obtained by cutting off 27 from the 61 rectangle and 16 from the 51 rectangle.

The median and quartiles have been stated to 1 decimal place in order to encourage the reader to draw a graph and work as accurately as possible. Let us now round them off to the nearest kg, since the original weights were given that way, and state

$$\text{median} = 76 \text{ kg},$$
$$\text{lower quartile} = 72 \text{ kg},$$
$$\text{upper quartile} = 81 \text{ kg}.$$

The median, percentiles and quartiles are discussed in more detail in §§ 20–3, 36 and 38.

19a. Probability graph paper. It is possible to obtain graph paper ruled in a special way called *Normal probability paper.* The divisions of the *ordinate scale* ('vertical' scale) of this graph paper are cumulative percentage frequencies spaced according to the Normal probability law. The divisions of the *abscissa scale* ('horizontal' scale) are ordinary units. This has the effect of pulling the Normal ogive from its characteristic shape into a straight line. In order to justify the statement that the distribution of weights of men *B* is approximately Normal it is first necessary to express the cumulative frequencies as percentages, as shown in table 2 B and then plot them on Normal probability paper as shown in fig. 9. We see then that the central 96 % of the distribution

TABLE 2B

Cumulative percentage frequencies of the weights of men B

Weight (kg)	Cumulative percentage frequency
Less than 59½	0·00
Less than 64½	1·95
Less than 69½	14·45
Less than 74½	38·28
Less than 79½	68·76
Less than 84½	88·67
Less than 89½	96·90
Less than 94½	98·06
Less than 99½	99·22
Less than 104½	100·00

Note that in this table 79½ kg is called the 68·76th *percentile.*
Alternatively 68·76 is called the *percentile rank* of 79½ kg.

gives a straight line with 2% at each end (the abnormal specimens) falling away from it. This straight line on the Normal probability paper indicates that the distribution is approximately Normal.

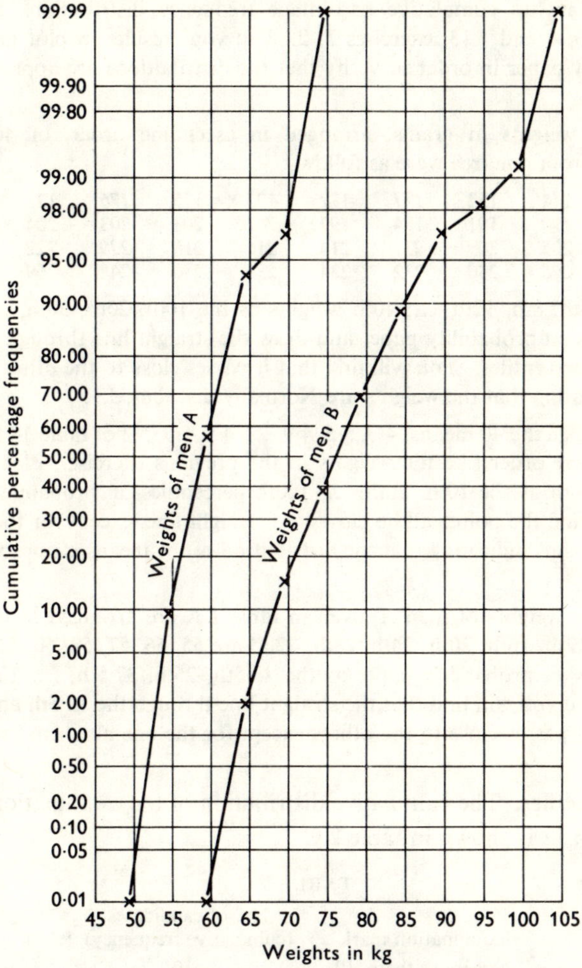

Fig. 9. The cumulative percentage frequency polygon or ogive plotted on Normal probability graph paper.

It is left as an exercise for the reader to obtain the cumulative percentage frequencies for the weights of men *A* which lead to the ogive also shown in fig. 9. Even from the small sample of 32 observations there is a strong suggestion by the graph that the weights of men *A* are also approximately Normal.

Further use can be made of the ogive drawn on Normal probability paper. This is described in § 59*a*.

19b. Exercises on probability paper.

1–4. Draw up cumulative percentage frequency distributions for § 11, exercises 3, 4 and § 13, exercises 1, 2. Use your results to plot graphs on probability paper in order to verify that the distributions are approximately Normal.

5. The weights in grams, arranged in ascending order, of 40 apples gathered from one tree were as follows:

120	144	152	*171*	172	173	175	*176*	178	181
185	*190*	191	194	197	*201*	203	203	204	*204*
205	207	209	*211*	214	216	218	*220*	222	224
226	*228*	230	232	234	*236*	240	248	269	*291*

Plot the 4th, 8th, 12th, ..., 40th weights as the 10th, 20th, 30th, ..., 100th percentiles on probability paper and draw the straight line through the 10th and 90th percentiles. You will find that it passes close to the other 8 points thus indicating that the weights are Normally distributed.

6. Arrange the 40 means, 4·3, 5·5, 4·9, ..., 4·5, 5·5, 5·2 of table 14 B, § 156, in ascending order like the weights in the previous exercise. Plot the 4th, 8th, ..., 40th as the 10th, 20th, ..., 100th percentiles on probability paper. You will find the points all lie close to a straight line, except for the highest mean 6·5, thus indicating that the distribution of the means is approximately Normal.

7. If the weights of men *A* given in table 1 A are arranged in order the 4th, 8th, 12th, 16th, 20th, 24th, 28th, 32nd are 55, 56, 57, 58, 60, 61, 62, 71. Plot these on probability paper as the 12·5th, 25th, 37·5th, ..., 100th percentiles and you will find that the straight line through the 12·5th and 87·5th percentiles passes close to the others except for the one abnormal weight of 71 kg.

20. Percentiles. The cumulative distribution of the examination marks given in § 12 is shown in table 2 C.

TABLE 2C

Examination mark	No. of candidates (cumulative frequency)
Not more than 10	10
Not more than 20	50
Not more than 30	130
Not more than 40	270
Not more than 50	440
Not more than 60	570
Not more than 70	670
Not more than 80	740
Not more than 90	780
Not more than 100	800

Note that the phrase 'Not more than 10' can be expressed in symbols as '$\leqslant 10$'.

The student should plot an accurate ogive of table 2 C in order to verify the answers to the following questions.

Question (i). *If the pass mark is 40, what percentage of the candidates pass the examination?*

The ogive shows that the number of candidates with not more than 39 is 256. Therefore 256 out of 800 fail. Hence 32% fail and 68% pass. Note that the examination mark 39 is called the *32nd percentile* because 32% of the candidates obtain 39 or less.

Question (ii). *If it is decided to allow 80% of the candidates to pass, what should the pass mark be?*

The ogive shows that 20%, i.e. 160, of the candidates have not more than 32 marks. Hence the pass mark should be 33. In this case 32 is the 20th percentile.

Question (iii). *If scholarships are awarded to the top 15% of the candidates, what would be the lowest mark to gain a scholarship?*

The ogive shows that 85%, i.e. 680, of the candidates have not more than 71 marks. Hence 72 is the lowest mark to gain a scholarship. In this example 71 is the 85th percentile.

If an ogive is drawn from a cumulative frequency distribution in which the phrase 'less than' is used instead of 'not more than' the arguments used in (i), (ii) and (iii) above must be modified.

21. The median. If a group of boys is arranged in order of weight, the weight of the middle boy is called the *median.* If, for example, the weights in kilos of 7 boys are

61, 63, 63, 64, 68, 71, 71,

the weight of the 4th boy, 64 kg, is the median.

Similarly, if the ages of 5 boys are

14 yr 3 mth, 14 yr 9 mth, 14 yr 9 mth,
14 yr 10 mth, 14 yr 11 mth,

the age of the middle boy, i.e. the 3rd boy, 14 yr 9 mth, is the median.

Further, if the heights in m of 10 boys are

1·40, 1·44, 1·48, 1·50, 1·51, 1·57, 1·60, 1·63, 1·64, 1·65,

the median is taken as $\frac{1}{2}(1{\cdot}51+1{\cdot}57) = 1{\cdot}54$ m because, due to the number of boys in the group being even, the 5th and 6th boys of heights 1·51 and 1·57 m are both in the middle. Thus, in this case, the median is defined as the arithmetic mean or average of the heights of the 5th and 6th boys, and it is known as the height of the 5·5th boy. Hence it may be stated in general that *if N measurements or observations are arranged in order of magnitude, the* MEDIAN *is the* $\frac{1}{2}(N+1)$ *the measurement or observation.*

If the weights of the 32 men *A* given in table 1 A are arranged in ascending order from 51 to 71 the median will be the weight of the $\frac{1}{2}(32+1)$th = the 16·5th weight. It will be found that the 16th and 17th men both weigh 57 kg. Hence the median is 57 kg. In table 1 B the 128th and 129th men both weigh 76 kg and this agrees exactly with the median predicted by the ogive in § 18.

22. The median of a large population. The marks of the 800 candidates given in table 2 C are an example of a *large population.* The median is the mark of the 400·5th candidate. The separate marks, however, are not available. Instead, the cumulative distribution is given and the ogive shows that 400·5 candidates get not more than 48 marks. It is fairly safe to assume, therefore, that the 400·5th candidate gets 48 marks and hence 48 is taken as the median. It will be realized that the difference between the 400·5th and the 400th is negligible. It cannot in fact be detected on the graph. Hence it should be noted that when the number of measurements *N* is large the median is taken as the $\frac{1}{2}Nth$ measurement. Alternatively, the median may be defined as the 50th percentile. It has half of the population above it and half below it. *It divides the area of the histogram into two equal parts.*

23. The quartiles. The *25th percentile* is called the *lower quartile* and the *75th percentile* the *upper quartile.* The ogive of table 2 C shows that 200 candidates have not more than 35 marks, while 600 candidates have not more than 63 marks.

Hence *the 25th percentile or lower quartile is 35 marks* and *the 75th percentile or upper quartile is 63 marks.*

The precise definition of the lower and upper quartiles is as follows: *If N measurements are arranged in order of magnitude, the* $\frac{1}{4}(N+1)th$ *and the* $\frac{3}{4}(N+1)th$ *are the* LOWER QUARTILE *and* UPPER QUARTILE *respectively.* It will be understood that $\frac{1}{4}(N+1)$th and $\frac{3}{4}(N+1)$th approximate to $\frac{1}{4}N$th and $\frac{3}{4}N$th respectively when *N* is large. Further, it will be realized that the lower quartile has one-quarter of the total population below it and three-quarters above it, while the upper quartile has three-quarters of the total population below it and one-quarter above it. *The median and quartiles together divide the area of the histogram into four equal parts.*

24. Exercises.

1. Arrange the examination marks of the 84 boys given in § 11, exercise 4, in order of magnitude and hence find the median and quartiles.

2. Use table 1 H to form a cumulative frequency distribution of the male population. Under 1, under 2, under 5, under 10, ..., under 100, would be the correct wording.

Draw the cumulative frequency curve or ogive and use it to estimate the percentage of the total male population (i) under 16, (ii) over 65, (iii) over 18 but under 20.

3. Form a cumulative frequency distribution of the girths of trees given in § 13, exercise 2. The appropriate wording in this case is not more than 0·6, not more than 0·8, etc.

Draw the ogive and use it to determine the median and quartiles.

4. Form a cumulative frequency distribution from the data of § 13, exercise 1. The wording suggested in this case is less than $54\frac{1}{2}$, less than $59\frac{1}{2}$, etc., because the weights have been given to the nearest kg.

Draw the ogive and determine the median and quartiles.

5. Distinguish between *median* and *mode*.

In a savings group there are 800 members, and the numbers of savings certificates held by them are shown in the following frequency table. Construct a table showing the cumulative frequency distribution and draw the graph of the ogive. Find (i) the mode, (ii) the median and (iii) the quartiles.

No. of certificates held	1–50	51–100	101–150	151–200	201–250
No. of members	5	10	30	60	100
No. of certificates held	251–300	301–350	351–400	401–450	451–500
No. of members	140	220	150	50	35

(London.)

6. The following table gives the frequency distribution of marks obtained by 130 candidates in each of two subjects A and B. Construct a table showing the cumulative frequency distributions in each subject and draw in one diagram the graphs of their ogives.

From your diagram, find

(*a*) the percentage number of candidates that fail in each subject if the pass mark in subject A is 55 and that in subject B is 35;

(*b*) the median mark in each subject.

Percentage	1–10	11–20	21–30	31–40	41–50
Subject A	0	0	1	3	6
Subject B	5	26	30	28	25
Percentage	51–60	61–70	71–80	81–90	91–100
Subject A	24	30	31	22	13
Subject B	9	5	0	1	1

(London.)

7. The following table gives the frequency distribution of the marks obtained in a test given to 300 candidates. Construct a table showing the cumulative frequency and draw a graph of the ogive. Obtain from the graph the median and the quartiles.

By extending the division of the distribution into percentiles, obtain also the mark not reached by the lowest 40% of the candidates and the percentage passing if the pass mark is 50.

0–10	11–20	21–30	31–40	41–50	51–60
10	10	20	20	25	25
61–70	71–80	81–90	91–100	101–110	111–120
40	55	45	20	20	10

(London.)

8. The table below shows the frequency distribution of marks of 800 candidates at an examination. Construct a table showing the cumulative frequency distribution and draw a graph of the ogive.

Use your ogive to determine:

(i) the median mark;

(ii) the percentage number of candidates that fails if the pass mark is 50;

(iii) the quartiles.

1–10	11–20	21–30	31–40	41–50	51–60	61–70	71–80	81–90	91–100
30	50	100	150	150	130	90	60	30	10

(London.)

9. Represent the following distribution in the form of an ogive or cumulative frequency curve and hence obtain estimates of the median and quartiles:

Daily wages (£)	No. of earners within given wage limits
1·00–	1
2·00–	7
3·00–	24
4·00–	36
5·00–	25
6·00–	6
7·00–	1
Total	100

(London.)

10. Construct an ogive or cumulative frequency curve for the following distribution and read off the median and quartiles:

Gross Margins of 180 Department Stores, 1927
(net sales over $1,000,000)

Gross margins as percentage of net sales	Frequency
18·5–	1
20·5–	1
22·5–	3
24·5–	1
26·5–	10
28·5–	21
30·5–	31
32·5–	61
34·5–	36
36·5–	10
38·5–	5

(London.)

11. *Chest Girth and Height of Males aged 20 examined by Medical Boards, Great Britain, 1939*

Chest girth (mm)	% cases	Height (mm)	% cases
Under 750	0·3	Under 1550	0·8
750–	2·3	1550–	3·3
800–	11·6	1600–	12·2
850–	30·4	1650–	25·4
900–	35·2	1700–	28·9
950–	16·1	1750–	19·6
1000–	3·4	1800–	7·4
1050 and over	0·7	1850 and over	2·4
Total	100·0	Total	100·0

Draw the ogives for each of the above distributions.

Note that, due to the frequencies being stated as percentages, the ogives are graphs of percentiles.

Determine the median and quartiles of each distribution. (London.)

12. The numbers of defective balls in equal batches of tennis balls produced by a company on 36 consecutive days are

3,	1,	2,	4,	0,	1,	2,	2,	3,	2,	1,	1,
5,	0,	1,	1,	1,	3,	2,	3,	4,	0,	1,	1,
2,	2,	3,	0,	3,	1,	2,	0,	1,	4,	3,	2.

Construct a frequency table showing for $x = 0, 1, 2, \ldots, 6$, the number n of batches having x defective balls. Find the mode and the median of this distribution and explain what these measures are. (W.A.E.C.)

3

Measures of location

25. The arithmetic mean of a set of numbers. The *arithmetic mean*, or more simply the *mean*, of the five numbers 23·46, 23·56, 23·67, 23·98, 22·43 is

$$\tfrac{1}{5}(23{\cdot}46+23{\cdot}56+23{\cdot}67+23{\cdot}98+22{\cdot}43) = \tfrac{1}{5}(117{\cdot}10) = 23{\cdot}42.$$

This number 23·42 is commonly called the *average* of the five numbers, but in the study of Statistics it is usually called the *arithmetic mean* or *mean.* The mean of a group of observations is a *measure of location* of the group. It is a *single number* which enables us to assess the position in which the group is *located* with respect to other groups. This single number is also called the *central tendency* of the group. Measures of location other than the mean are discussed later in this chapter.

The above calculation can be simplified in the following way. First subtract 23 from each of the numbers and write then as 0·46, 0·56, 0·67, 0·98, −0·57. Next calculate their mean 0·42. Finally, replace the 23 and thus obtain 23·42. This process is described as 'working with 23 as origin', the number 23 being called the 'arbitrary origin', 'starting point' or 'working zero'.

An even greater simplification is to write the numbers as 46, 56, 67, 98, −57, ignoring for the time being the fact that they are *hundredths.* Their mean is calculated as 42. The 42 is then written in hundredths as 0·42 and finally the 23 is replaced giving 23·42. This is described as 'working with 23 as origin, and 0·01 as unit'.

Second example. By working with 15 years as origin and 1 month as unit, calculate, to the nearest month, the mean age of eight boys whose ages are 15 yr 3 mth, 15 yr 9 mth, 14 yr 10 mth, 15 yr 0 mth, 15 yr 7 mth, 14 yr 4 mth, 15 yr 5 mth, 16 yr 0 mth.

Calculation. The mean of 3, 9, −2, 0, 7, −8, 5, 12 is

$$\tfrac{1}{8}(3+9-2+0+7-8+5+12) = 3\tfrac{1}{4} \text{ months},$$

and hence the mean age = 15 yr $3\frac{1}{4}$ mth
= 15 yr 3 mth (to the nearest month).

26. Exercises.

1. Working with 980 as origin and 0·1 as unit calculate the mean of 980·8, 981·1, 980·7, 980·3, 981·8, 982·5.

2. A player makes the following scores when competing in ten games: 47, 41, 50, 39, 45, 48, 42, 32, 60, 20. Calculate his mean score per game by taking 40 as origin.

3. Calculate, correct to 1 decimal place, the mean of the following percentage relatives: 100·2, 101·3, 101·8, 98·4, 100·5, 100·2, 101·6, 97·5, 99·8.

27. When some of the observations are repeated. If the weights, in kilos, of twelve men are given as 73, 72, 72, 71, 71, 71, 70, 70, 69, 69, 69, 69, the mean weight is

$$70+\tfrac{1}{12}(3+2+2+1+1+1+0+0-1-1-1-1) = 70{\cdot}5 \text{ kg.}$$

The weights might have been given in a frequency table:

Weight (kg)	73	72	71	70	69
Frequency	1	2	3	2	4

The mean weight might then have been calculated as

$$70+\tfrac{1}{12}\{3+2(2)+3(1)+2(0)+4(-1)\} = 70{\cdot}5 \text{ kg.}$$

It is convenient to tabulate the last calculation as given in table 3 A.

TABLE 3A

Weight taking 70 as origin (kg) x	Frequency f	Frequency × weight fx
3	1	3
2	2	4
1	3	3
0	2	0
–1	4	–4
Total	12	6

$$\text{Mean weight} = 70+\left(\frac{\text{Sum of the } fx \text{ column}}{\text{Sum of the } f \text{ column}}\right) \text{kg.}$$
$$= 70{\cdot}5 \text{ kg.}$$

28. The mean of a frequency distribution. The method of the last paragraph can be applied to the frequency distribution in table 1 D to calculate the mean weight of men *B* provided we assume that the 5 men whose weights are between $59\frac{1}{2}$ and $64\frac{1}{2}$ kg are all 62 kg (the mid-value of the group), the 32 men whose weights are between $64\frac{1}{2}$ and $69\frac{1}{2}$ are all 67 and so on. This assumption is justified because the errors it introduces in the individual weights are positive and negative and tend to cancel each other particularly when the frequency of each group is large. If the *mid-interval weight*, 77 kg, of the modal group is taken as the arbitrary origin or working zero and *5 kg are taken as one unit*, the calculation is as shown in table 3 B.

Note that an alternative way of describing the values x is, 'deviations of the mid-interval weights from 77 kg taking 5 kg as unit' since, for example, 87 kg deviates from 77 kg by $+2$ 5 kg units while 72 kg deviates from 77 kg by -1 5 kg units.

The mean, 76·7 kg, is written to 1 decimal place because it is customary to state means to 1 significant figure more than the original observations. The mean is close to the median in this case because the distribution is approximately Normal.

TABLE 3B

Method of calculating the mean frequency distribution

Weights of men B (kg)	Frequency (f)	Mid-interval weights in kg of the groups in the first column	Mid-interval weights 77 kg as origin 5 kg as unit x	fx	
59½–64½	5	62	-3		-15
64½–69½	32	67	-2		-64
69½–74½	61	72	-1		-61
74½–79½	78	77	0	0	
79½–84½	51	82	1	51	
84½–89½	21	87	2	42	
89½–94½	3	92	3	9	
94½–99½	3	97	4	12	
99½–104½	2	102	5	10	
Total	256	—	Total	124	-140

As x is measured in 5 kg units, fx is in 5 kg units and so it is necessary to multiply $\left(\dfrac{\text{Sum of the } fx \text{ column}}{\text{Sum of the } f \text{ column}}\right)$ by 5 to bring it to 1 kg units before adding it to the 77 kg which was taken as origin. Hence the mean weight of men B is

$$77+\frac{5(124-140)}{256} = 76{\cdot}7 \text{ kg}.$$

29. Exercises.

1. Rewrite the calculation shown in table 3 B taking 72 kg as origin instead of 77 kg and verify that an identical result is obtained.

2. Calculate from table 1 C the mean weight of men A. Use 57 as origin and 5 as unit.

30. The use of Σ. The phrase 'the *sum* of the *fx* column' is used so often in statistical calculations that it is convenient to represent it by Σfx, The symbol Σ, pronounced 'sigma', is the capital *S* of the Greek alphabet. Similarly Σf is short for 'the sum of the *f* column'.

31. A second example using mid-interval values. The method of § 28 can be applied to the data of § 12 to calculate the mean examination mark of the 800 candidates as shown in table 3 C.

Note that the way in which the groups of a frequency distribution are stated must always be carefully studied to decide the mid-interval values. An error in the mid-interval values causes an equal error in the mean. Examination marks grouped 0–9, 10–19, etc., have mid-interval values $4\frac{1}{2}$, $14\frac{1}{2}$, etc., but weights of men in kg grouped 50–, 60–, etc., would have mid-interval weights 55, 65, etc., because '50–' means 50–60 but not including 60. The variation in weights is *continuous* with the dividing line drawn at 60, whereas the variation in the marks is *discrete* with 9 belonging to one group and 10 to the next, there being no possible mark between.

TABLE 3C

Examination mark	No. of candidates f	Mid-interval marks of the groups of the first column	Mid-interval marks $45\frac{1}{2}$ as origin 10 as unit x	fx	
1–10	10	$5\frac{1}{2}$	−4		−40
11–20	40	$15\frac{1}{2}$	−3		−120
21–30	80	$25\frac{1}{2}$	−2		−160
31–40	140	$35\frac{1}{2}$	−1		−140
41–50	170	$45\frac{1}{2}$	0	0	
51–60	130	$55\frac{1}{2}$	1	130	
61–70	100	$65\frac{1}{2}$	2	200	
71–80	70	$75\frac{1}{2}$	3	210	
81–90	40	$85\frac{1}{2}$	4	160	
91–100	20	$95\frac{1}{2}$	5	100	
Total	800	—	Total	800	−460

$$\text{Mean examination mark} = 45\tfrac{1}{2} + \frac{10\Sigma fx}{\Sigma f}$$

$$= 45\tfrac{1}{2} + \frac{10(800-460)}{800}$$

$$= 49\tfrac{3}{4}.$$

32. Exercises.

1. Rewrite the calculation shown in § 28 taking 82 kg as origin and 5 kg as unit and verify that an identical result is obtained.

2. Rewrite the calculation shown in § 31 taking $55\frac{1}{2}$ as origin and 10 as unit and verify that an identical result is obtained.

3. Taking a working zero of 77 kg and 5 kg as unit, calculate the mean weight of the 100 men given in § 13, exercise 1.

4. By taking an arbitrary origin in the 1·0–1·2 group calculate the mean girth of the trees given in § 13, exercise 2.

5. The 'recovery time' of an aircraft is the time that elapses between its arrival at an airport and its being made ready to take-off again.

Calculate the mean recovery time of the 100 aircraft shown in the following distribution:

Recovery time (min)	5–10	10–15	15–20	20–25
No. of aircraft	4	22	28	18
Recovery time (min)	25–30	30–35	35–40	40–45
No. of aircraft	12	9	4	3

6. Use the data of § 24, exercise 5, to obtain the mean holding of certificates.

7. Use the data of § 24, exercise 9, to obtain the mean daily wage.

8. Use the data of § 24, exercise 11, to calculate the mean chest girth and height.

9. The table below gives the frequency distribution of the ages (to the nearest month) of 200 boys in a secondary school.

Age in months	Number of boys
141–145	20
146–150	28
151–155	90
156–160	40
161–165	22

Construct a histogram and use it to estimate the mode of this distribution. Estimate also the mean of the distribution. Which of these two measures do you consider as more representative of the given distribution? State your reasons. (W.A.E.C.)

33. Unequal grouping. The calculation of the mean age of the male population of table 1 H involves dealing with unequal class intervals. The mid-interval ages for under 1, 1 but under 2, 60–99 are respectively $\frac{1}{2}$, $1\frac{1}{2}$, 80. The mid-interval of the 2–4 group is $3\frac{1}{2}$ because this group may include boys who are only 1 day short of being 5. Similarly, the mid-interval of the 5–9 group is $7\frac{1}{2}$. The full calculation is shown in table 3 D.

TABLE 3D

Calculation of the mean of a frequency distribution with unequal grouping

Mid-interval age (years)	Percentage of total male population f	Mid-interval age $17\frac{1}{2}$ as origin 5 years as unit x	fx	
$\frac{1}{2}$	1·8	−3·4		− 6·12
$1\frac{1}{2}$	1·7	−3·2		− 5·44
$3\frac{1}{2}$	4·8	−2·8		−13·44
$7\frac{1}{2}$	7·5	−2		−15·00
$12\frac{1}{2}$	8·0	−1		− 8·00
$17\frac{1}{2}$	8·5	0	0	
30	24·0	2·5	60·00	
50	26·0	6·5	169·00	
80	17·7	12·5	221·225	
Total	100·0	—	450·25	−48·00

$$\text{Mean age} = 17\tfrac{1}{2} + \frac{5(402{\cdot}25)}{100}$$

$$= 37{\cdot}6 \text{ years.}$$

34. Discussion of measures of location. The *arithmetic mean*, *mode* and *median* are all measures of location. So are the *geometric mean* and *harmonic mean*, which are also described and discussed in this chapter. A measure of location has already been described as a numerical characteristic of a group which enables us to assess the position in which the group stands with respect to other groups. The group under consideration, for example, might be a variety of potato whose position with respect to other varieties is being assessed by the average weight of potatoes produced per root, or it might be a workshop whose position with respect to other workshops is being assessed by the average number of articles produced per week. It might, indeed, be an individual such as a cow whose position in the herd is assessed by its average milk yield, or, of course, a games player whose position among other players is assessed by his average score per game.

35. The arithmetic mean. The arithmetic mean is the best known and most useful measure of location. It is fundamental to nearly all statistical theory. By multiplying the mean by the total frequency the aggregate for the group is obtained. Means of separate groups can easily be manipulated to obtain the mean of the combined group. For example, if the mean age of 15 boys is m_1 years and the mean age of 14 girls is m_2 years, the mean age of the 29 boys and girls is $\frac{1}{29}(15m_1 + 14m_2)$

years. Sometimes abnormal individuals among the observations can have a rather exaggerated effect on the mean, and this is the chief criticism that is usually levelled against it. Suppose, for example, the Savings Collection at a school was approximately £5 each week for 40 weeks but £210 on one particular week when a special campaign was organized.

$$\text{The mean weekly collection} = £\tfrac{1}{41}(40 \times 5 + 210) = £10.$$

Thus the one abnormal week has made the mean equal to double the normal collection.

36. The median. The median can easily be picked out when the individuals are ranged in order, and it provides an *average specimen* for examination. In a form, the boy of median age can be interviewed, the physique of the boy of median weight can be examined, and the exercise books of the boy with the median mark can be studied carefully. The median is entirely unaffected by abnormal individuals. It is also unaffected if the observations of a group are *transformed* by some process such as *taking logarithms* or *squaring*. It is, however, unsuitable for work demanding mathematical manipulation.

37. The mode. The mode seems most useful to manufacturers of articles by sizes. It is obviously useful for the shoemaker to know the modal size of shoe and for the cycle manufacturer to know the modal size of frame. The mode is not easy to determine with precision when the observations fall into groups. Indeed, it is quite possible that no two observations may be alike. Like the median, the mode is unaffected by abnormal individuals and transformations but it is unsuitable for mathematical manipulation.

38. Relation between mean, median and mode. An interesting method of estimating approximately the mode is by the formula

$$\text{mean} - \text{mode} = 3(\text{mean} - \text{median})$$

which applies to distributions which have one hump only.

In §§ 22 and 31, for example, we found that the median and mean marks of the 800 candidates were 48 and $49\frac{3}{4}$ respectively. Hence the mode $= 49\frac{3}{4} - 3(49\frac{3}{4} - 48) = 44\frac{1}{2}$, which is approximately the estimation of § 12.

39. The geometric mean. The *geometric mean* of n numbers is the nth root of their product. Thus the geometric mean of the five numbers 1, 2, 3, 4, 5 is $\sqrt[5]{(1.2.3.4.5)} = 2{\cdot}605$ by logarithms. Note that the geometric mean is less than the arithmetic mean. It is not influenced by abnormal individuals to the same extent. Further, the geometric mean of two ratios p/a and q/b is equal to the ratio of the geometric mean $\sqrt{(pq)}$ to the geometric mean $\sqrt{(ab)}$. This can easily be verified for any four positive numbers a, b, p, q. This property, which does not hold for the arithmetic mean, makes the geometric mean more satisfactory for averaging percentage relatives (see chapter 12).

40. The harmonic mean. The *reciprocal* of the *harmonic mean* of n numbers is the *arithmetic mean of their reciprocals.* Thus the harmonic means of 1, 2, 3, 4, 5 is given by

$$\begin{aligned} \frac{1}{\text{harmonic mean}} &= \frac{1}{5}\left(\frac{1}{1}+\frac{1}{2}+\frac{1}{3}+\frac{1}{4}+\frac{1}{5}\right) \\ &= \tfrac{1}{5}(1+0{\cdot}5+0{\cdot}3333+0{\cdot}25+0{\cdot}2) \\ &= 0{\cdot}4567. \end{aligned}$$

Hence the harmonic mean $= 2{\cdot}19$.

It should be noted that the harmonic mean is less than the geometric mean and that it might be alternatively defined as the *reciprocal of the arithmetic mean of the reciprocals.*

The use of the harmonic mean is illustrated by the following example. If a boat travels s km upstream at u km/h and then returns the s km downstream at v km/h, the $2s$ km have been covered in $\left(\frac{s}{u}+\frac{s}{v}\right)$ hours, and hence the average speed for the whole journey is

$$\frac{2s}{\frac{s}{u}+\frac{s}{v}} = \frac{1}{\frac{1}{2}\left(\frac{1}{u}+\frac{1}{v}\right)}$$ km/h, which is the harmonic mean of u and v.

41. Exercises.

1. An aircraft travels 500 km at 200 km/h and returns over the same route at 250 km/h. Calculate its average speed for the whole journey.

2. A car does its first 10,000 km at 12 km to the litre of petrol and its next 10,000 km at 10 km to the litre. What is its average petrol consumption for the whole 20,000 km?

4

Measures of dispersion

42. Dispersion or variability. Suppose that the points scored by two players in seven separate games are as shown in the following table:

	Scores for seven separate games							Mean score
Player *A*	20	65	0	50	100	35	80	50
Player *B*	40	55	35	50	65	45	60	50

It is clear that, although the mean score of each player is the same, the scores of *A* are much more dispersed, scattered or spread about the mean than those of *B*. The scores of 100 and 80 made by *A* are better than any made by *B*, the scores of 0 and 20 are worse. The performances of *A* are more variable than those of *B*, and indeed, less reliable. The purpose of this chapter is to investigate the methods by which the *dispersion* or *variability* of groups of observations can be compared.

43. Examples.

1. A well-known sugar refiner uses machines which pack automatically 1 kg cartons of sugar. When these machines are in operation the cartons emerge from them in a continuous stream ready for sale. In order to check that a particular machine is giving correct weight sample cartons are occasionally selected from it and weighed accurately. The results of checking two machines are as follows:

	Accurate weights in kilos of eleven cartons					
Machine *A*	1·017	1·051	1·078	0·996	1·033	1·059
Machine *B*	0·995	1·009	1·028	1·036	1·000	1·017
Machine *A*	1·082	1·014	1·040	1·072	0·998	
Machine *B*	1·027	1·045	1·006	1·018	1·039	

Note that the dispersion of *A* is greater than that of *B* because the weights delivered by machine *A range* from 0·996 to 1·082 kg, while those delivered by machine *B range* from 0·995 to 1·045 kg.

2. A firm which manufactures lead-covered submarine cables checked the thickness of the cover on two of its cables by taking measurements in ten places:

	Ten measurements, in mm, of the thickness of the cover									
Cable *A*	7·4	7·6	7·8	7·0	7·2	7·3	7·5	7·7	7·9	7·1
Cable *B*	7·0	7·2	7·3	7·4	7·2	7·2	7·4	7·1	7·2	7·3

Here the measurements *A range* from 7·0 to 7·9 while measurements *B range* from 7·0 to 7·4, and hence the dispersion of *A* is greater than that of *B*.

3. *English and Mathematics marks. Frequency distribution of 1000 candidates*

Examination mark	0–9	10–19	20–29	30–39	40–49
No. of candidates in English	1	3	19	131	346
No. of candidates in Mathematics	20	41	98	149	192
Examination mark	50–59	60–69	70–79	80–89	90–99
No. of candidates in English	340	134	20	4	2
No. of candidates in Mathematics	197	152	97	44	10

Since the number of candidates with very low and very high marks is greater in Mathematics than in English, the dispersion of the Mathematics marks may be taken to be greater than the dispersion of the English marks. If the frequency polygons of the two distributions are drawn they show quite clearly the difference in dispersion. In fig. 10 the

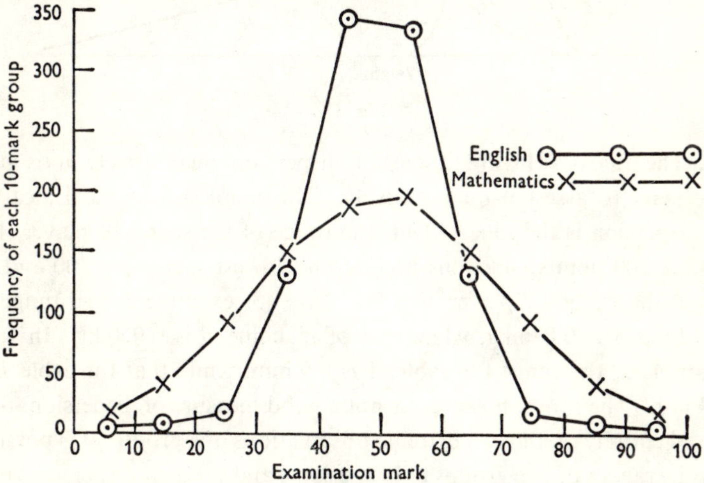

Fig. 10

tall narrow frequency polygon for the English marks indicates that a great number of candidates are packed close to the mean, while the short wide frequency polygon for the Mathematics marks indicates that the candidates are widely scattered away from the mean. Hence the dispersion is small for English and large for Mathematics.

4. A study of the weights of 100 cats and 100 dogs led to the following statements:

(i) The average weight of the dogs was greater than that of the cats.

(ii) The variability in weight was greater for the dogs.

(iii) The smallest dog was about the same weight as the smallest cat.

(iv) The biggest cats were smaller than the average dogs.

(v) The distribution of the weights of the cats was approximately normal.

(vi) The distribution of the weights of the dogs was skewed positively.

It is interesting to express the above statements diagrammatically by means of frequency curves as shown in fig. 11.

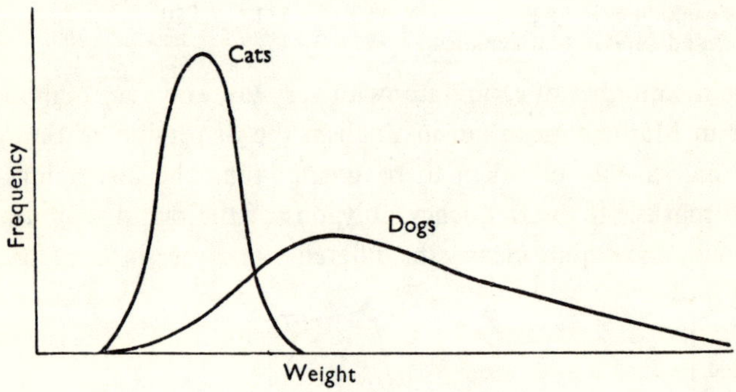

Fig. 11

44. The range. Having discussed dispersion qualitatively it is now necessary to assess it quantitatively. The simplest numerical measure of dispersion is the *range*. Thus the range of the scores of player A in § 42 is 100 points (since his highest and lowest scores are 100 and 0), while the range of player B is 30. In § 43, example 1, the range of machine A is 0·086 kg, while that of machine B is 0·050 kg. In § 43, example 2, the range for cable A is 0·9 mm, while that for cable B is 0·4 mm. The range, however, is not a good measure of dispersion. It is based entirely on the two extreme observations of a group. It is possible for the ranges of two groups to be the same and yet their dispersions may be different. The examination marks of eleven candidates in Art, for example, might be

0, 25, 35, 40, 45, 50, 55, 60, 65, 75, 100.

while their marks in French might be

0, 10, 20, 30, 40, 50, 60, 70, 80, 90, 100.

Here, except for the 0 and the 100, the Art marks are clustered more closely to the mean than the French. Hence the dispersion of the Art marks is less than the dispersion of the French marks, although the range is the same for each. Moreover, the range generally depends on the size of the sample, i.e. on the number of observations. The range of scores in twenty games, for example, is likely to be greater than the range for seven games.

45. Interpercentile ranges. A second measure of dispersion is the range between the 10th percentile and the 90th percentile. If the ogives are drawn for § 43, example 3, the following results will be obtained from them:

	English marks	Mathematics marks
90th percentile	$63\frac{1}{2}$	74
10th percentile	35	23
10–90 percentile range	$28\frac{1}{2}$	51

This indicates that the dispersion of the Mathematics marks is almost twice that of the English marks. The 10–90 *percentile range* does not depend on the extreme observations and it does make some use of the whole group of observations, but it does not lend itself to further mathematical treatment in statistical analysis.

46. The semi-interquartile range or quartile deviation. A more frequently used measure of dispersion than the 10–90 percentile range is the *semi-interquartile range*. This, as its name implies, is half the range between the quartiles. Quite often it is called the *semi-quartile range* or *quartile deviation*. Thus

$$\text{Semi-interquartile range or quartile deviation} = \frac{\text{Upper quartile} - \text{Lower quartile}}{2}.$$

From the ogive of § 43, example 3, the following results can be obtained:

	English marks	Mathematics marks
Upper quartile	56	$62\frac{1}{2}$
Lower quartile	42	35
Interquartile range	14	$27\frac{1}{2}$
Semi-interquartile range	7	$13\frac{3}{4}$

Thus the dispersion of the Mathematics marks is again indicated as almost twice that of the English marks. As the semi-interquartile range ignores completely the top quarter and bottom quarter of the observations and is based entirely on the range of the middle half, it is not a satisfactory measure of the dispersion of the whole group. Nor does it lend itself to further mathematical treatment.

47. The mean deviation. A measure of dispersion which makes use of all the observations of a group is the *mean deviation.* This is calculated for the machine A of § 43, example 1, in table 4A in which

(i) the first column contains the weights of the cartons together with their total and mean;

(ii) the second column contains the differences between the weights and their mean. These differences are called the *deviations from the mean;*

(iii) the third column contains these *deviations all taken as positive.*

TABLE 4A

	Weights of cartons (kg)	Deviations of the weights from the mean 1·040 kg d kg	Deviations all taken as positive $\|d\|$ kg
	1·017	−0·023	0·023
	1·051	+0·011	0·011
	1·078	+0·038	0·038
	0·996	−0·044	0·044
	1·033	−0·007	0·007
	1·059	+0·019	0·019
	1·082	+0·042	0·042
	1·014	−0·026	0·026
	1·040	0·000	0·000
	1·072	+0·032	0·032
	0·998	−0·042	0·042
Total	11·440	0·000	0·284
Mean	1·040	—	0·026

The mean deviation = the mean of the deviations all taken as positive
= 0·026 kg.

Note that $|d|$ means the magnitude of d without regard to its sign. It is called the *modulus* of d.

The student will find it a useful exercise to calculate by the same method the mean deviation for machine B. The result obtained, 0·014 kg, indicates that the dispersion or variability of machine A is almost twice that of machine B.

In calculating the mean deviation, the deviations may be measured from the MEAN, MEDIAN *or* MODE. *They are really best measured from the median, however, because the* MEAN DEVIATION FROM THE MEDIAN *is a* MINIMUM.

When the variability or dispersion of one group is being compared with that of another group the mean deviation must clearly be measured from the same average for each group.

48. Exercises.

1. Compare the variability in the thickness of the cover of the submarine cables of § 43, example 2, by calculating the mean deviation from the median in each case. Note that the median for cable *A* is 7·45 mm, while that for cable *B* is 7·2 mm.

2. Calculate the mean deviation from the mean of the six values given in § 26, exercise 1.

3. Calculate the mean deviation from the mean of the ten scores of § 26, exercise 2.

4. Determine the median and the mean deviation from the median of the nine percentage relatives of § 26, exercise 3.

5. The following are the index numbers of production in the textile and food industries for 10 years between 1924 and 1934, excluding 1926:

Textiles	Food
100	100
122	97
120	103
112	103
114	108
90	109
89	113
101	112
109	110
112	115

Compare the variability of production in the two industries by calculating the mean deviation from the median for each set of index numbers.

(J.M.B.)

6. Compare the variability in price of beef and mutton by calculating the mean deviation from the median for the prices shown in the following table:

	Price	
Year	Beef	Mutton
1925	80	106
1926	74	89
1927	70	86
1928	74	92
1929	71	89
1930	73	92
1931	67	79
1932	65	63
1933	61	69
1934	58	74
1935	54	75
1936	54	73
1937	61	78
1938	62	62

(J.M.B.)

49. Mean deviation of a frequency distribution. Table 4 B shows the calculation of the mean deviation from the median of the English marks given in § 43, example 3. The median is $49\frac{1}{2}$ because 500 candidates have 49 or less and 500 candidates have 50 or more.

TABLE 4B

Mid-interval mark	Frequency f	Deviations of mid-interval mark from the median $49\frac{1}{2}$ d	Deviations all taken as positive $\lvert d\rvert$	$f\lvert d\rvert$
$4\frac{1}{2}$	1	−45	45	45
$14\frac{1}{2}$	3	−35	35	105
$24\frac{1}{2}$	19	−25	25	475
$34\frac{1}{2}$	131	−15	15	1965
$44\frac{1}{2}$	346	− 5	5	1730
$54\frac{1}{2}$	340	+ 5	5	1700
$64\frac{1}{2}$	134	+15	15	2010
$74\frac{1}{2}$	20	+25	25	500
$84\frac{1}{2}$	4	+35	35	140
$94\frac{1}{2}$	2	+45	45	90
Total	1000	—	—	8760

$$\text{Mean deviation} = \frac{\text{Sum of the } f\lvert d\rvert \text{ column}}{\text{Sum of the } f \text{ column}}$$

$$= \frac{\Sigma f\lvert d\rvert}{\Sigma f}$$

$$= 8{\cdot}76.$$

The student should calculate by the same method the mean deviation of the Mathematics marks. It is an unusual coincidence that the median is the same as for the English marks. The value obtained for the mean deviation should be 15·66, thus indicating that the dispersion of the Mathematics marks is almost twice that of the English marks.

50. Exercises.

Calculate the mean deviation from the median for the following:

1. The number of certificates of § 24, exercise 5.
2. The subjects *A* and *B* of § 24, exercise 6.
3. The examination marks of § 24, exercise 7.
4. The daily wages of § 24, exercise 9.
5. The gross margin percentages of § 24, exercise 10.

51. The standard deviation. The most satisfactory and most widely used measure of dispersion is the *standard deviation.* This is calculated for a small number of observations as follows:

(i) Write down the *squares of the deviations* from the mean. These will, of course, all be positive.

(ii) Calculate the *mean of these squares of deviations.*

(iii) The standard deviation is the *square root* of *the mean of the squares of deviations.*

The standard deviation is always measured from the MEAN *and never from the median or mode.*

For the machine *A* of § 43, example 1, the calculation of the standard deviation might be tabulated as shown in table 4 C.

TABLE 4C

	Weights of cartons (kg)	Deviations of the weights from the mean 1·040 kg $d$kg	Squares of the deviations d^2kg^2
	1·017	−0·023	0·000529
	1·051	+0·011	0·000121
	1·078	+0·038	0·001444
	0·996	−0·044	0·001936
	1·033	−0·007	0·000049
	1·059	+0·019	0·000361
	1·082	+0·042	0·001764
	1·014	−0·026	0·000676
	1·040	0·000	0·000000
	1·072	+0·032	0·001024
	0·998	−0·042	0·001764
Total	11·440	0·000	0·009668
Mean	1·040	—	0·000879

$$\text{Standard deviation} = \sqrt{\left(\frac{\Sigma d^2}{11}\right)}\ \text{kg}$$
$$= 0{\cdot}0296\ \text{kg.}$$

A similar calculation for machine *B* gives the standard deviation 0·0157 kg, thus indicating that the dispersion of machine *A* is almost twice that of machine *B*.

52. Avoiding the decimals. If the deviations in the last calculation had been written with 0·001 kg as unit the decimals would have been avoided. The total for the d^2 column would then be 9668, its mean 879 and the final step in the calculation would be

$$\text{Standard deviation} = 0{\cdot}001\sqrt{\left(\frac{\Sigma d^2}{11}\right)}$$
$$= 0{\cdot}001 \times 29{\cdot}6\ \text{kg}$$
$$= 0{\cdot}0296\ \text{kg.}$$

The factor 0·001 is introduced to bring back the units to kg. The weights of the cartons, in the first place, could have been converted to grams. It is, however, a useful exercise to retain the original units by means of the factor 0·001.

53. Using an arbitrary origin. Much arithmetical labour can be avoided by using an arbitrary origin in the calculation of the standard deviation. Consider the thickness of the cover of cable *A* given in § 43, example 2. Their mean is 7·45 mm, giving deviations −0·05, +0·15, +0·35, etc., and even if the decimals are avoided by working with 0·01 mm as unit the squares are still rather cumbersome. Table 4 D shows the method of easing the burden by working with one of the central observations 7·5 mm as origin.

TABLE 4D

Thickness of cover (mm)	Thickness with 7·5 mm as origin and 0·1 mm as unit x	Squares of deviations from 7·5 mm 0·1 mm as unit x^2
7·4	−1	1
7·6	+1	1
7·8	+3	9
7·0	−5	25
7·2	−3	9
7·3	−2	4
7·5	0	0
7·7	+2	4
7·9	+4	16
7·1	−4	16
Total	−5	85

By the method of § 25

$$\text{the mean} = 7{\cdot}5 + \frac{0{\cdot}1\times(-5)}{10}\text{ mm}$$

$$= 7{\cdot}45\text{ mm.}$$

Now $\Sigma x^2 = 85$ is the sum of the squares of deviations from 7·5 mm and therefore

$0{\cdot}1\sqrt{\left(\frac{\Sigma x^2}{10}\right)}$ is *not* the standard deviation. It can be proved, however, that

$$\text{the standard deviation} = 0{\cdot}1\sqrt{\left\{\frac{\Sigma x^2}{10} - \left(\frac{\Sigma x}{10}\right)^2\right\}},$$

the term $-\left(\frac{\Sigma x}{10}\right)^2$ being the correction necessary because the deviations are from 7·5 mm instead of 7·45 mm Hence

$$\begin{aligned}\text{the standard deviation} &= 0{\cdot}1\sqrt{\{8{\cdot}5-(-0{\cdot}5)^2\}}\text{ mm}\\ &= 0{\cdot}1\sqrt{\{8{\cdot}5-0{\cdot}25\}}\text{ mm}\\ &= 0{\cdot}2872\text{ mm.}\end{aligned}$$

It can be verified that, if deviations are measured in mm from the true mean 7·45 mm, the sum of the deviations is zero, the sum of the squares of deviations is 0·825 $(\text{mm})^2$ and hence the result obtained is the same.

The student will find it a useful exercise to calculate by the same method the standard deviation for cable *B*. By working with 7·2 mm as origin and 0·1 mm as unit it will be found that *zero* occurs four times in the *x* column. This makes the arithmetic very easy. The results obtained should be

Mean = 7·23 mm
Standard deviation = 0·119 mm.

54. Mathematical justification of the method. The method of § 53 will now be justified by replacing the values $-1, +1, +3, \ldots, -4$ of table 4 D by the 10 observations $x_1, x_2, x_3, \ldots, x_{10}$. Let $\bar{x}$ be their mean so that the deviation of x_1 from $\bar{x}$ is $x_1-\bar{x}$. The following table shows the necessary steps in the calculation, from first principles, of the standard deviation:

	10 observations x	Deviations of the observations from the mean x d	Squares of deviations d^2
	x_1	$x_1-\bar{x}$	$x_1^2-2x_1\bar{x}+\bar{x}^2$
	x_2	$x_2-\bar{x}$	$x_2^2-2x_2\bar{x}+\bar{x}^2$
	x_3	$x_3-\bar{x}$	$x_3^2-2x_3\bar{x}+\bar{x}^2$
	$\vdots$	$\vdots$	$\vdots$
	x_{10}	$x_{10}-\bar{x}$	$x_{10}^2-2x_{10}\bar{x}+\bar{x}^2$
Total	Σx	Total	$\Sigma x^2-2\bar{x}\Sigma x+10\bar{x}^2$

Note that the sum of the d^2 column, Σd^2, is

$$(x_1^2+x_2^2+x_3^2+\ldots+x_{10}^2)-2\bar{x}(x_1+x_2+x_3+\ldots+x_{10}) + (\bar{x}^2+\bar{x}^2+\bar{x}^2+\ldots+\bar{x}^2),$$

which can be written, as shown,

$$\Sigma x^2-2\bar{x}\Sigma x+10\bar{x}^2.$$

Now the mean $\bar{x}=\dfrac{\Sigma x}{10}$ and, by definition, the standard deviation is

$$\sqrt{\left\{\frac{\Sigma d^2}{10}\right\}}=\sqrt{\left\{\frac{\Sigma x^2-2\bar{x}\Sigma x+10\bar{x}^2}{10}\right\}}$$

$$=\sqrt{\left\{\frac{\Sigma x^2}{10}-2\left(\frac{\Sigma x}{10}\right)\left(\frac{\Sigma x}{10}\right)+\left(\frac{\Sigma x}{10}\right)^2\right\}}$$

$$=\sqrt{\left\{\frac{\Sigma x^2}{10}-\left(\frac{\Sigma x}{10}\right)^2\right\}}.$$

It should be noted that

(i) *the change of unit has not been mentioned because it is merely an artifice for removing decimals at the beginning of the calculation and putting them back at the end;*

(ii) *the change of origin has not been mentioned because the deviations of the observations from the mean are independent of the origin.*

The general result for N observations is:

$$\textit{the standard deviation} = \sqrt{\left\{\frac{\Sigma x^2}{N} - \left(\frac{\Sigma x}{N}\right)^2\right\}}.$$

It can be proved by replacing 10 by N in the preceding argument.

55. Exercises.

1. Working with 981 as origin and 0·1 as unit calculate the standard deviation of the six values given in § 26, exercise 1.

2. By taking 40 as origin calculate the standard deviation of the ten scores of § 26, exercise 2.

3. By taking 100 as origin and 0·1 as unit calculate the standard deviation of the 9 percentage relatives given in § 26, exercise 3.

4. Use the data of § 48, exercise 5, to calculate the standard deviation (i) for textiles, (ii) for food, and thus compare the variability of production in the two industries.

5. Use the data of § 48, exercise 6, to calculate the standard deviation of (i) the beef prices, (ii) the mutton prices.

56. The standard deviation of a frequency distribution. The standard deviation of the English marks given in § 43, example 3, will now be calculated. The method is an extension of that shown in § 28 for calculating the mean by working with the mid-interval value of the modal class as arbitrary origin. It is shown in table 4 E.

Note that Σx^2, Σx and N of § 53 become Σfx^2, Σfx and Σf respectively when the standard deviation of a frequency distribution is being calculated. Also, it is easier to calculate the values in the fx^2 column by multiplying the fx value by the x value than by multiplying the f value by the square of the x value.

The student should now calculate by this method the mean and standard deviation of the Mathematics marks of the same example. The results obtained should be:

Mean = 49·2
Standard deviation = 19·1.

TABLE 4E

Mid-interval mark	Frequency f	Deviations of the mid-interval mark from $44\frac{1}{2}$, 10 as unit x	fx		fx^2
$4\frac{1}{2}$	1	-4		-4	16
$14\frac{1}{2}$	3	-3		-9	27
$24\frac{1}{2}$	19	-2		-38	76
$34\frac{1}{2}$	131	-1		-131	131
$44\frac{1}{2}$	346	0	0		0
$54\frac{1}{2}$	340	1	340		340
$64\frac{1}{2}$	134	2	268		536
$74\frac{1}{2}$	20	3	60		180
$84\frac{1}{2}$	4	4	16		64
$94\frac{1}{2}$	2	5	10		50
Total	1000	—	694	-182	1420

(i) The mean examination mark $= 44\frac{1}{2} + \frac{10\Sigma fx}{\Sigma f}$

$$= 44\frac{1}{2} + \frac{10 \times 512}{1000}$$

$$= 49{\cdot}62.$$

(ii) The standard deviation $= 10 \times \sqrt{\left\{\frac{\Sigma fx^2}{\Sigma f} - \left(\frac{\Sigma fx}{\Sigma f}\right)^2\right\}}$

$$= 10 \times \sqrt{\{1{\cdot}420 - (0{\cdot}512)^2\}}$$

$$= 10{\cdot}76.$$

57. A comparison of the various measures of dispersion. The chief objection to the mean deviation is that it ignores plus and minus signs. This means that it does not lend itself to mathematical treatment. Nor is it easy to use an arbitrary origin for its calculation. The standard deviation may, at first, seem more complicated to the beginner, but it is easily calculated by working with an arbitrary origin and, like the mean, it is fundamental to nearly all statistical theory. Table 4 F gives a summary of the advantages and limitations of the various measures of dispersion. *All are measured in the same units as the given observations.*

58. The variance. The variance is the square of the standard deviation. If the final square root is not extracted in the calculation of the standard deviation the result obtained is then the variance. It is a measure of dispersion but it has the disadvantage of being measured in units which are the square of those of the given observations. If the heights of a group of men are given in mm, the standard deviation is in mm but the variance is in $(mm)^2$. In statistical theory the variance and standard deviation have a fundamental importance which may be compared with *the square on the hypotenuse* and *the length of the hypotenuse* in geometrical theory. This is discussed further in §§ 162–8.

TABLE 4F

Range	Simple idea	Based entirely on the two extremes	Unsuitable for further mathematical treatment
Interpercentile ranges	Easily found from the ogive	Ignores entirely the extremes	Unsuitable for further mathematical treatment
Semi-interquartile range or quartile deviation	Easily found from the ogive	Ignores the top and bottom quarters and is based entirely on the middle half of the observations	Unsuitable for further mathematical treatment
Mean deviation	Easily understood. Working with an arbitrary origin not easy; calculation can, therefore, be tedious	Makes use of all observations but ignores signs	Unsuitable for further mathematical treatment
Standard deviation	Easily calculated by working with an arbitrary origin	Makes use of all observations; does not ignore signs	Fundamental to nearly all advanced work

59. The coefficient of variation. When dealing with positive observations, two groups which have completely different mean values or are measured in completely different units can be compared by calculating the *coefficient of variation* in each case. This is defined by

$$\text{coefficient of variation} = \frac{\text{standard deviation}}{\text{mean}},$$

or alternatively by

$$\text{coefficient of variation} = \frac{100 \times \text{standard deviation}}{\text{mean}}\%,$$

i.e. the standard deviation is expressed either as a fraction or as a percentage of the mean. The coefficient of variation is independent of the units in which the mean and standard deviation are measured. The standard deviation is an *absolute* measure, the coefficient of variation a *relative* measure of dispersion.

Example 1. The machine *A* of § 43, example 1, packs cartons of sugar of mean weight 1·040 kg and standard deviation 0·0296 kg. A similar machine packs cartons with mean weight 2·070 kg and standard deviation 0·035 kg. Hence

the coefficient of variation of the 1 kg machine = 0·0296/1·040
= 0·028,

and the coefficient of variation of the 2 kg machine = 0·035/2·070
= 0·017.

Thus the 2 kg machine is less variable than the 1 kg machine.

Example 2. The mean height of a group of men is 1780 mm with standard deviation 33 mm, while the mean weight is 76 kg with standard deviation 3·3 kg. Thus

the coefficient of variation of the heights = 33/1780
= 0·019,

and the coefficient of variation of the weights = 3·3/76
= 0·043.

Hence the weights are more dispersed than the heights.

59a. Estimation of the mean and standard deviation by probability paper. *If a distribution is Normal*

the mean = the 50th percentile

the standard deviation = the difference between the 50th and 84th percentiles.

This fact is explained fully in Chapter 7. It is convenient to use it here to show that probability paper can be used to estimate the mean and the standard deviation of a distribution which is approximately Normal.

Figure 9 shows that the distributions of the weights of men *A* and men *B* are approximately Normal. From it we can read off the following results:

	Weights of men *A* (kg)	Weights of men *B* (kg)
50th percentile (the mean)	59·0	76·0
84th percentile	62·5	82·6
Standard deviation	3·5	6·6

If instead of joining the points to get the graph we had drawn one straight line through the middle of the points (as in fig. 23) we should get slightly different and probably better estimates.

The following results are obtained by the calculation method of § 56:

	Weights of men *A* (kg)	Weights of men *B* (kg)
Mean	59·2	76·7
Standard deviation	4·1	6·9

60. Exercises.

1. Calculate the standard deviation of the distribution given in § 24, exercise 5. Take $275\frac{1}{2}$ as origin and work with 50 as unit.

2. Compare the variability of subjects *A* and *B* of § 24, exercise 6, by calculating the standard deviation of each. For *A*, work with $65\frac{1}{2}$ as origin and 10 as unit and for *B*, work with $35\frac{1}{2}$ as origin and 10 as unit.

3. Taking £4·50 as origin and £1 as unit calculate the standard deviation of the distribution of § 24, exercise 9.

4. Calculate the standard deviation of the distribution of § 24, exercise 10.

5. By calculating the coefficient of variation of the chest girths and of the heights given in § 24, exercise 11, determine which of the two physical characteristics is the more dispersed.

6. The mean *recovery time* for the aircraft of a certain type *A* is 18 min with a standard deviation of 3 min, while for a different type *B* the mean is 50 min with a standard deviation of 9 min. By calculating the coefficient of variation in each case determine which type has the greatest variability in recovery time.

7. The mean *length of life* of a certain type of television receiver tube is 1600 h with a standard deviation of 250 h, while the mean length of life of the valves for the same receivers is 1000 h with standard deviation 200 h. By calculating the coefficient of variation in each case determine which has the greater variability of life.

8–12. Obtain estimates of the mean and standard deviation for exercises 1–5 above by using probability paper as shown in § 59*a*. Compare the graphical estimates with the results obtained by calculation.

5

Probability

61. Casting a die. When an ordinary die is cast, each of the numbers *one*, *two*, *three*, *four*, *five*, *six* has an *equal chance* of falling uppermost. There is, therefore, 1 *chance out of* 6 of obtaining a *six* and we say that

$$\textit{the probability of throwing a six} = \tfrac{1}{6}.$$

62. Playing cards. An ordinary pack of 52 playing cards contains 4 *aces* and hence, if we select at random 1 card from a well-shuffled pack, there are 4 chances out of 52 of it being an *ace*. Thus

$$\textit{the probability of selecting an ace} = \tfrac{4}{52} = \tfrac{1}{13}.$$

Further,

$$\textit{the probability of selecting the ace of spades} = \tfrac{1}{52}.$$

63. Tossing a coin. When a coin is tossed it may fall either as a *head* or a *tail* and thus

$$\textit{the probability of a head} = \tfrac{1}{2}.$$

64. Definition of probability. Sections 61, 62 and 63 are simple illustrations of the definition of probability which is as follows: *When all the equally possible occurrences have been enumerated the probability of a particular event is the ratio of the number of ways in which the particular event may occur to the total number of possible occurrences.*

65. The meaning of probability. When we say that the probability of throwing a *six* is $\frac{1}{6}$ we do not mean that exactly 1 throw out of every 6 will produce a *six*. If the student actually throws a die 60 times it is *likely* that he will obtain approximately 10 *sixes*. If he throws one 600 times a *reasonable estimate* of the number of *sixes* he can expect is 100 but it is very unlikely that it will be exactly 100. The following table gives the results of an actual experiment in which a die was cast 648 times. The frequency of each number approximates to 108 which is what we should expect since the probability of each number is $\frac{1}{6}$:

Number thrown	1	2	3	4	5	6
Frequency	96	98	117	130	107	100

66. Success and failure. If we regard the throwing of a *six* as a *success* and the throwing of any other number as a *failure*, the probability of success is $\frac{1}{6}$ and the probability of failure is $\frac{5}{6}$. It is customary to use the letter p for the probability of success and the letter q for the probability of failure. Thus $p = \frac{1}{6}$ and $q = \frac{5}{6}$ and $q = 1-p$.

In § 62, if the selecting of an *ace* is a success, $p = \frac{1}{13}$ and $q = \frac{12}{13}$ while in § 63, if a *head* is a success, $p = \frac{1}{2}$ and $q = \frac{1}{2}$ also.

67. Addition rule. Suppose a student casts a die and that he is to receive a prize if he obtains a *six* OR a *one*. The probability of him winning a prize is

$$\tfrac{1}{6} + \tfrac{1}{6} = \tfrac{1}{3}.$$

This example illustrates the *addition rule* which states: *The probability that* ONE OR OTHER *of several* MUTUALLY EXCLUSIVE *events shall occur is the* SUM *of the probabilities of the separate events.* Note particularly that the throwing of a *six* excludes the possibility of a *one* and the throwing of a *one* excludes the possibility of a *six*.

68. Multiplication rule. Suppose the student casts a die, selects a card from a well-shuffled pack and tosses a coin and that he is to receive a prize only if he obtains a *six* AND an *ace* AND a *head*. The probability of his success is then

$$\tfrac{1}{6} \times \tfrac{1}{13} \times \tfrac{1}{2} = \tfrac{1}{156}.$$

This is an example of the *multiplication rule* which states: *If the probabilities of several independent events are $p_1, p_2, \ldots, p_n$, the probability that* ALL *will take place is the* PRODUCT $p_1 p_2 \cdots p_n$.

69. The use of $q = 1-p$. Suppose the student of § 68 is to receive a prize provided

(i) he throws a *six* but neither selects an *ace* nor tosses a *head*,
OR (ii) he selects an *ace* but neither throws a *six* nor tosses a *head*,
OR (iii) he tosses a *head* but neither throws a *six* nor selects an *ace*.

Consider case (i). The probability that he throws a *six* is $\frac{1}{6}$, the probability that he does not select an *ace* is $1-\frac{1}{13} = \frac{12}{13}$ and the probability that he does not toss a *head* is $\frac{1}{2}$. Thus the probability of case (i) occurring is $\frac{1}{6} \times \frac{12}{13} \times \frac{1}{2}$. Similarly the probabilities of cases (ii) and (iii) are $\frac{1}{13} \times \frac{5}{6} \times \frac{1}{2}$ and $\frac{1}{2} \times \frac{5}{6} \times \frac{12}{13}$ respectively and (i), (ii) and (iii) are mutually exclusive events. Hence the total probability of the boy's success is

$$\tfrac{1}{6} \times \tfrac{12}{13} \times \tfrac{1}{2} + \tfrac{1}{13} \times \tfrac{5}{6} \times \tfrac{1}{2} + \tfrac{1}{2} \times \tfrac{5}{6} \times \tfrac{12}{13} = \tfrac{77}{156}.$$

This example illustrates the use of the important principle: *If the probabilities of several independent events are $p_1, p_2, p_3, \ldots, p_n$, the probability that the first takes place and that the others do not is*

$$p_1(1-p_2)(1-p_3)\ldots(1-p_n).$$

70. Examples.

1. *If three dice are thrown together, what is the probability of obtaining*

(i) *3 sixes;*
(ii) *1, and only 1, six;*
(iii) *at least 1 six?*

(i) By the multiplication rule, the probability of 3 *sixes* is $(\frac{1}{6})^3 = \frac{1}{216}$.

(ii) By the method of § 69, the first die may be a *six* with the second and third not *or* the second a *six* and the first and third not *or* the third a *six* with the first and second not. Thus the total probability is $3(\frac{1}{6} \times \frac{5}{6} \times \frac{5}{6}) = \frac{25}{72}$.

(iii) The probability of obtaining at least 1 *six*

$= 1 -$ the probability of obtaining no *sixes*

$= 1 - (\frac{5}{6})^3$

$= \frac{91}{216}$.

2. *If two cards are drawn from a well-shuffled pack of 52 playing cards, what is the probability that*

(i) *they are both aces;*
(ii) *neither of them is an ace;*
(iii) *at least one of them is an ace?*

(i) The probability that the first is an ace is $\frac{4}{52}$.

If the first is an ace there are only 3 aces left in the remaining 51 cards and the probability that the second is an ace is $\frac{3}{51}$. Thus the probability that both are aces is $(\frac{4}{52}) \times (\frac{3}{51}) = \frac{1}{221}$.

(ii) The probability that neither of them is an ace is $(\frac{48}{52}) \times (\frac{47}{51}) = \frac{188}{221}$.

(iii) The probability that at least one of them is an ace is

$$1 - (\frac{188}{221}) = \frac{33}{221}.$$

3. *A box contains ten batteries all apparently sound, although four of them are actually defective. Find the chance that, if two of the batteries are tested, they are both defective.*

The probability that the first one tested is defective is $\frac{4}{10}$. If the first battery tested is defective, the probability that the second is also defective is $\frac{3}{9}$. Hence the probability that both are defective is $\frac{4}{10} \times \frac{3}{9} = \frac{2}{15}$.

71. Exercises.

1. When three marksmen take part in a shooting contest their chances of hitting the target are $\frac{1}{2}$, $\frac{1}{3}$ and $\frac{1}{4}$. Calculate the chance that one, and only one, bullet will hit the target if all three men fire at it simultaneously. (J.M.B.)

2. The independent probabilities that three components of a television set will need replacing within a year are $\frac{1}{10}$, $\frac{1}{12}$ and $\frac{1}{15}$. Calculate the probability that (i) at least one component, (ii) one and only one component, will need replacing. (J.M.B.)

3. Five per cent of a large consignment of eggs are bad. Find the probability of getting at least one bad egg in a random sample of dozen. (J.M.B.)

4. A bag contains 7 black balls and 3 white balls. If they are drawn one by one from the bag, find the probability of drawing first a black, then a white and so on alternately until only black balls remain.

5. Three of five dice are each numbered in the normal way, but the fourth is numbered 1, 2, 3, 6, 5, 6 and the fifth 1, 6, 3, 6, 5, 6. If the five dice are thrown together, calculate the probability of turning up (i) five 6's, (ii) at least four 6's, (iii) at least one 6. (J.M.B.)

6. In a class of 24 boys and 16 girls, 15 of the boys and 12 of the girls study Biology.

(*a*) A boy and a girl are selected at random from the class. Find the probability that only one of the two studies Biology.

(*b*) One name is selected at random from a list of all those who do not study Biology. Find the probability that the person selected is a boy.

(*c*) One name is selected at random from a list of all students in the class. Find the probability that the person selected is a boy who does not study Biology. (W.A.E.C.)

6

The Binomial distribution

72. Casting two dice together. Suppose two dice are thrown together and that the throwing of a *six* is regarded as a *success* but the throwing of any other number as a *failure*. For each die, the probability of success $p = \frac{1}{6}$, the probability of failure $q = \frac{5}{6}$ and, by an argument similar to § 70, for the two dice thrown together

the probability of no successes, $P(0) = q^2 = 25/36$,
the probability of 1 success, $P(1) = 2qp = 10/36$,
the probability of 2 successes, $P(2) = p^2 = 1/36$

and the total probability

$$\begin{aligned} P(0)+P(1)+P(2) &= q^2+2qp+p^2 \\ &= (q+p)^2 \\ &= 1. \end{aligned}$$

Experiment. The reader should actually throw 2 dice 216 times and form a *relative frequency distribution* to compare with the above *probability distribution.*

73. Casting three dice. The throwing of 3 dice together has already been discussed in § 70. Hence, again assuming that the probability of success $p = \frac{1}{6}$ and the probability of failure $q = \frac{5}{6}$, we may state for the 3 dice thrown together

the probability of no successes, $P(0) = q^3 = 125/216$,
the probability of 1 success, $P(1) = 3q^2p = 75/216$,
the probability of 2 successes, $P(2) = 3qp^2 = 15/216$,
the probability of 3 successes, $P(3) = p^3 = 1/216$,

and the total probability

$$\begin{aligned} P(0)+P(1)+P(2)+P(3) &= q^3+3q^2p+3qp^2+p^3 \\ &= (q+p)^4 \\ &= 1. \end{aligned}$$

Experiment. Throw 3 dice 216 times, form a relative frequency distribution and compare it with the above probability distribution.

74. Casting four or more dice. The arguments of §§ 72 and 73 may be continued indefinitely. For four dice thrown together we find that

$$\begin{aligned} P(0)+P(1)+P(2)+P(3)+P(4) &= q^4+4q^3p+\frac{4\times3}{1\times2}q^2p^2+\frac{4\times3\times2}{1\times2\times3}qp^3+p^4 \\ &= q^4+4q^3p+6q^2p^2+4qp^3+p^4 \\ &= (q+p)^4 \\ &= 1, \end{aligned}$$

for five dice thrown together we find that

$$\begin{aligned} P(0)&+P(1)+P(2)+P(3)+P(4)+P(5) \\ &= q^5+5q^4p+\frac{5\times4}{1\times2}q^3p^2+\frac{5\times4\times3}{1\times2\times3}q^2p^3+\ldots+p^5 \\ &= q^5+5q^4p+10q^3p^2+10q^2p^3+5qp^4+p^5 \\ &= (q+p)^5 \\ &= 1, \end{aligned}$$

and so on for 6 dice, 7 dice etc.

Experiments similar to those suggested in §§ 72 and 73 should be carried out with 4 dice, 5 dice, etc. to compare the probability distributions with the experimental frequency distributions.

75. The Binomial distribution. The throwing of dice is easy to understand and easy to carry out experimentally. For that reason it has been made the basis of our discussion. In the sections which follow examples such as the tossing of coins and sampling in industry and market research will be described. The throwing of 6 dice or the tossing of 6 coins may be generalized as 6 *independent trials* of the same event. The complete generalization is to state the case of n independent trials as follows:

Suppose the probability of an event occurring at a single trial is p and the probability of it not occurring is q then $p+q=1$ and the probabilities

$$P(0),\ P(1),\ P(2),\ P(3),\ \ldots,\ P(n)$$

of $0, 1, 2, \ldots, n$ occurrences of the event in n independent trials are given by the terms of the *Binomial expansion*

$$(q+p)^n = q^n+nq^{n-1}p+\frac{n(n-1)}{1\,.\,2}q^{n-2}p^2+\frac{n(n-1)(n-2)}{1\,.\,2\,.\,3}q^{n-3}p^3+\ldots+p^n.$$

and an appropriate name for the probabilities is, therefore, a *Binomial distribution.*

Table 6 A gives the first 6 probabilities of Binomial distributions in which the number of independent trials $n = 2, 3, 4, 5, 6, 8, 10, 12, 20, 50$.

TABLE 6A

The Binomial distribution of probabilities when the probability of success is p and the probability of failure q

No. of independent trials n	Probability of no successes $P(0)$	1 success $P(1)$	2 successes $P(2)$	3 successes $P(3)$	4 successes $P(4)$	5 successes $P(5)$
2	q^2	$2qp$	p^2			
3	q^3	$3q^2p$	$3qp^2$	p^3		
4	q^4	$4q^3p$	$6q^2p^2$	$4qp^3$	p^4	
5	q^5	$5q^4p$	$10q^3p^2$	$10q^2p^3$	$5qp^4$	p^5
6	q^6	$6q^5p$	$15q^4p^2$	$20q^3p^3$	$15q^2p^4$	$6qp^5$
8	q^8	$8q^7p$	$28q^6p^2$	$56q^5p^3$	$70q^4p^4$	$56q^3p^5$
10	q^{10}	$10q^9p$	$45q^8p^2$	$120q^7p^3$	$210q^6p^4$	$252q^5p^5$
12	q^{12}	$12q^{11}p$	$66q^{10}p^2$	$220q^9p^3$	$495q^8p^4$	$792q^7p^5$
20	q^{20}	$20q^{19}p$	$190q^{18}p^2$	$1140q^{17}p^3$	$4845q^{16}p^4$	$15{,}504q^{15}p^5$
50	q^{50}	$50q^{49}p$	$1225q^{48}p^2$	$19{,}600q^{47}p^3$	$230{,}300q^{46}p^4$	$2{,}118{,}760q^{45}p^5$

76. Casting six dice. It is now possible to confirm the probabilities stated in § 8 for six dice thrown together. The information, including the result of an actual experiment of 216 throws, is shown in table 6 B.

TABLE 6B

Six dice thrown together

Number of sixes	Probability	Probable frequency for 216 throws 216 $P(x)$	Observed frequency in 216 throws
0	$P(0) = \left(\frac{5}{6}\right)^6$	72·34	77
1	$P(1) = 6\left(\frac{5}{6}\right)^5\left(\frac{1}{6}\right)$	86·81	84
2	$P(2) = 15\left(\frac{5}{6}\right)^4\left(\frac{1}{6}\right)^2$	43·40	34
3	$P(3) = 20\left(\frac{5}{6}\right)^3\left(\frac{1}{6}\right)^3$	11·57	19
4	$P(4) = 15\left(\frac{5}{6}\right)^2\left(\frac{1}{6}\right)^4$	1·74	7
5	$P(5) = 6\left(\frac{5}{6}\right)\left(\frac{1}{6}\right)^5$	0·14	0
6	$P(6) = \left(\frac{1}{6}\right)^6$	0·00	0
	Total	216·00	216

77. The notation $\binom{n}{r}$. The Binomial expansion of $(q+p)^n$ is often written

$$q^n + \binom{n}{1} q^{n-1} p + \binom{n}{2} q^{n-2} p^2 + \binom{n}{3} q^{n-3} p^3 + \ldots + p^n,$$

the symbols $\binom{n}{1}, \binom{n}{2}, \binom{n}{3}, \ldots$ being a convenient way of respectively representing

$$\frac{n}{1}, \quad \frac{n(n-1)}{1\,.\,2}, \quad \frac{n(n-1)(n-2)}{1\,.\,2\,.\,3}, \ldots$$

Thus, the first 6 terms of $(q+p)^7$ are written

$$(q+p)^7 = q^7 + \binom{7}{1} q^6 p + \binom{7}{2} q^5 p^2 + \binom{7}{3} q^4 p^3 + \binom{7}{4} q^3 p^4 + \binom{7}{5} q^2 p^5 + \ldots + p^7,$$

by which we mean

$$(q+p)^7 = q^7 + \frac{7}{1}q^6 p + \frac{7\times 6}{1\times 2}q^5 p^2 + \frac{7\times 6\times 5}{1\times 2\times 3}q^4 p^3 + \frac{7\times 6\times 5\times 4}{1\times 2\times 3\times 4}q^3 p^4$$

$$+\frac{7\times 6\times 5\times 4\times 3}{1\times 2\times 3\times 4\times 5}q^2 p^5 + \ldots + p^7$$

$$= q^7 + 7q^6 p + 21q^5 p^2 + 35q^4 p^3 + 35q^3 p^4 + 21q^2 p^5 + \ldots + p^7.$$

78. Tossing five coins. The terms of the binomial expansion

$$(q+p)^5 = q^5 + \binom{5}{1} q^4 p + \binom{5}{2} q^3 p^2 + \binom{5}{3} q^2 p^3 + \binom{5}{4} qp^4 + p^5,$$

where $q = \frac{1}{2}$ and $p = \frac{1}{2}$, give the probabilities of obtaining 0, 1, 2, 3, 4, 5 *heads* when 5 coins are tossed together. These probabilities together with the expected frequencies for a total of 832 tosses and the observed frequencies of an actual experiment are displayed in table 6 C. The student will find it interesting to carry out the experiment himself but not necessarily for such a large number of tosses.

TABLE 6C

Five coins tossed together

No. of heads x	Probability $P(x)$	Expected frequency for 832 tosses $832 . P(x)$	Observed frequency of an actual experiment
0	$P(0) = \left(\frac{1}{2}\right)^5$	26	25
1	$P(1) = \binom{5}{1}\left(\frac{1}{2}\right)^4\left(\frac{1}{2}\right)$	130	123
2	$P(2) = \binom{5}{2}\left(\frac{1}{2}\right)^3\left(\frac{1}{2}\right)^2$	260	246
3	$P(3) = \binom{5}{3}\left(\frac{1}{2}\right)^2\left(\frac{1}{2}\right)^3$	260	264
4	$P(4) = \binom{5}{4}\left(\frac{1}{2}\right)\left(\frac{1}{2}\right)^4$	130	146
5	$P(5) = \left(\frac{1}{2}\right)^5$	26	28
Total	1	832	832

79. Exercise.

Calculate the probabilities of obtaining 0, 1, 2, 3, 4 *heads* when 4 coins are tossed together. Estimate the frequencies that can be expected if the 4 coins are tossed together 40 times and test by experiment.

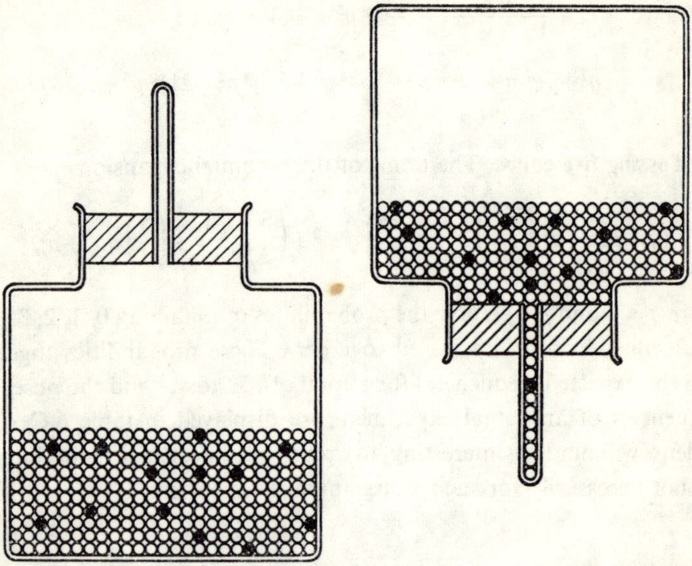

Fig. 12. A sampling-bottle. *Upright position*; the bottle contains a large number of small steel balls 5% of which are painted black; a hollow glass tube sealed at the outer end and open at the inner end passes through the cork. *Inverted position*; after being well shaken the bottle is turned over and some of the steel balls fall into the hollow glass tube, which is just long enough to allow 10 balls to be visible below the cork.

80. The sampling-bottle. The *sampling-bottle* shown in fig. 12 contains a large number of small steel balls 5% of which are painted black. If one ball is taken at random from the bottle the probability, p, that it is black is $\frac{1}{20}$ or 0·05 and the probability, q, that it is not black is $\frac{19}{20}$ or 0·95. Suppose a random sample of 10 balls is taken from the bottle. The probabilities that the sample contains 0, 1, 2, etc., black balls are given by the terms of the binomial expansion

$$(q+p)^{10} = q^{10} + \binom{10}{1} q^9 p + \binom{10}{2} q^8 p^2 + \ldots,$$

where $q = 0{\cdot}95$ and $p = 0{\cdot}05$. These probabilities are shown in table 6 D. The separate probabilities of 3, 4, 5, etc., black balls appearing in any random sample are so small that they have been combined as *the probability of* 3 *or more*, $P(x \geqslant 3)$, where

$$P(x \geqslant 3) = 1 - \{P(0) + P(1) + P(2)\}.$$

It will be appreciated that the hollow glass tube passing through the cork is a very convenient arrangement for drawing random samples from the bottle. The process of shaking the bottle, turning it over, counting the number of black balls visible below the cork, is so quick that it can be repeated a hundred times in a few minutes.

81. Exercises.

1. Draw up a table similar to 6 D for a sampling-bottle if 10% of the balls contained in it are black and the glass tube is just long enough to draw random samples of five.

2. Draw up a table similar to 6 D for a sampling-bottle if $2\frac{1}{2}$% of the balls contained in it are black and the glass tube is long enough to draw samples of twenty.

TABLE 6D

The sampling bottle of fig. 12

No. of black balls in the random sample of 10 x	Probability $P(x)$	Expected frequency for 100 samples $100.P(x)$
0	$P(0) = (0{\cdot}95)^{10} = 0{\cdot}598$	59·8
1	$P(1) = \binom{10}{1}(0{\cdot}95)^9(0{\cdot}05) = 0{\cdot}315$	31·5
2	$P(2) = \binom{10}{2}(0{\cdot}95)^9(0{\cdot}05)^2 = 0{\cdot}075$	7·5
3 or more	$P(x \geqslant 3) = 1 - \{P(0) + P(1) + P(2)\} = 0{\cdot}012$	1·2
Total	1	100·0

82. Sampling in industry. The experiments carried out with the sampling-bottle lead directly to the method of sampling as it is applied in the testing of mass-produced articles which are either accepted or rejected according as to whether or not they satisfy specification requirements. If, for example, 5% of the screws manufactured by a certain process are rejected because they fail to satisfy tolerance requirements, table 6 D gives the probabilities of a random sample of 10 screws containing 0, 1, 2, 3 or more rejects. Also, if 100 samples are inspected the frequencies for 0, 1, 2, 3 or more rejects per sample should approximate fairly closely to the expected frequencies of table 6 D, any large divergence signifying that the proportion of rejects in the bulk differs significantly from 5%. The examples and exercises which follow will serve to make clear the important practical applications of the Binomial distribution.

83. Examples.

1. (i) *In a large lot of electric-light bulbs 5% of the bulbs are defective. Calculate the probability that a random sample of 20 will contain at most 2 defective bulbs.*

(ii) *One third of the lots presented for inspection have 5% defective, the rest 10% defective. If a lot is rejected when a random sample of 20 taken from it contains more than 2 defective bulbs, find the proportion of lots which are rejected.*

(i) Using the terms of the binomial expansion

$$(q+p)^{20} = q^{20} + \binom{20}{1} q^{19} p + \binom{20}{2} q^{18} p^2 + \dots$$

with $p = 0{\cdot}05$ and $q = 0{\cdot}95$ it is convenient to tabulate the calculation in a form similar to table 6 D, as shown in table 6 E.

Thus, the probability that a random sample of 20 will contain at most 2 defective bulbs is 0·9236. This means that for lots which have 5% defective about 92% of the samples will have at most 2 defective bulbs and 8% of the samples will have more than 2 defective bulbs.

TABLE 6E

No. of defective bulbs in random sample of 20 x	Probability $P(x)$
0	$P(0) = (0{\cdot}95)^{20} = 0{\cdot}3581$
1	$P(1) = \binom{20}{1}(0{\cdot}95)^{19}(0{\cdot}05) = 0{\cdot}3770$
2	$P(2) = \binom{20}{2}(0{\cdot}95)^{18}(0{\cdot}00)^2 = 0{\cdot}1885$
At most 2	$P(x \leqslant 2) = P(0)+P(1)+P(2) = 0{\cdot}9236$

TABLE 6F

No. of defective bulbs in random sample of 20 x	Probability $P(x)$
0	$P(0) = (0{\cdot}9)^{20} = 0{\cdot}1213$
1	$P(1) = \binom{20}{1}(0{\cdot}9)^{19}(0{\cdot}1) = 0{\cdot}2696$
2	$P(2) = \binom{20}{2}(0{\cdot}9)^{18}(0{\cdot}1)^2 = 0{\cdot}2846$
At most 2	$P(x \leqslant 2) = P(0)+P(1)+P(2) = 0{\cdot}6755$

(ii) Table 6 F gives values for lots which have 10% defective $p = 0{\cdot}1$ and $q = 0{\cdot}9$.

Thus, for lots which have 10% defective about 68% of the samples will have at most 2 defective bulbs and 32% of the samples will have more than 3 defective bulbs.

Let us suppose that 300 lots are presented for inspection, 100 of them having 5% defective and 200 having 10% defective. Then 8 of the 100 and 2×32 of the 200 will be rejected.

Thus 72 of the 300 lots will be rejected.

This means that 24% of all the lots presented for inspection will be rejected.

2. *Assuming that the probability that any one of 8 telephone lines is engaged at an instant is $\frac{1}{4}$, calculate the probability that*

(i) *at least one of the lines is engaged,*

(ii) *all 8 lines are engaged.*

What is the most probable number of lines engaged at any instant and what is the probability that this number of lines is engaged?

The simplest method of calculation is by table 6 G.

TABLE 6G

No. of lines engaged x	Probability $P(x)$
0	$P(0) = (\frac{3}{4})^8 = 0{\cdot}1002$
1	$P(1) = \binom{8}{1}\left(\frac{3}{4}\right)^7\left(\frac{1}{4}\right) = 0{\cdot}2669$
2	$P(2) = \binom{8}{2}\left(\frac{3}{4}\right)^6\left(\frac{1}{4}\right)^2 = 0{\cdot}3114$
3	$P(3) = \binom{8}{3}\left(\frac{3}{4}\right)^5\left(\frac{1}{4}\right)^3 = 0{\cdot}2076$
$\vdots$	$\vdots$
8	$P(8) = (\frac{1}{4})^8 = 0{\cdot}00001524$

The probability that at least one line is engaged is

$$\begin{aligned} P(1)+P(2)+P(3)+\ldots+P(8) &= 1-P(0) \\ &= 0{\cdot}8998, \end{aligned}$$

which is approximately $\frac{9}{10}$.

The probability that all the lines are engaged is 0·00001524. The most probable number of lines engaged at any instant is 2 and the probability of this number of lines being engaged is 0·3114 (rather less than $\frac{1}{3}$).

A more direct way of finding the most probable number is as follows: The probabilities $P(0)$, $P(1)$, $P(2)$, etc., continue to increase provided

$$\frac{P(x+1)}{P(x)} \geqslant 1.$$

Now

$$\frac{P(x+1)}{P(x)} = \frac{\binom{8}{x+1}\left(\frac{3}{4}\right)^{7-x}\left(\frac{1}{4}\right)^{x+1}}{\binom{8}{x}\left(\frac{3}{4}\right)^{8-x}\left(\frac{1}{4}\right)^{x}}$$

$$= \left\{\frac{8!}{(x+1)!(7-x)!}\bigg/\frac{8}{x!(8-x)!}\right\}\frac{(\frac{1}{4})}{(\frac{3}{4})}$$

$$= \frac{(8-x)}{3(x+1)}.$$

Thus $\dfrac{P(x+1)}{P(x)} \geqslant 1$ provided $(8-x) \geqslant 3(x+1)$.

This gives $x \leqslant \frac{5}{4}$ and hence $P(2)$ is the greatest probability.

84. Exercises.

1. In the manufacture of screws by a certain process it was found that 8% of the screws were rejected because they failed to satisfy tolerance requirements. What was the probability that a sample of 10 screws contained (i) exactly 2, (ii) not more than 2 rejects?

2. Fifty samples of 20 are drawn from a large bulk in which the proportion of defective items is 12%. Calculate the number of samples that can be expected to include 0, 1, 2, etc., defectives.

3. If on the average rain falls on 12 days in every 30 find the probability (i) that the first 3 days of a given week will be fine and the remainder wet; (ii) that rain will fall on just 3 days of a given week. (London.)

4. A marksman's chance of hitting a target with each of his shots is $\frac{2}{5}$. If he fires 6 shots calculate his chance of (i) exactly 3 hits, (ii) at least 3 hits.

5. The probability that a tennis player will serve an 'ace' is $\frac{1}{3}$. What is the probability that she will serve exactly 4 aces out of 7 services?

85. Verification that the mean of the Binomial distribution is *np*.

(i) From table 6 B, the mean number of sixes per throw,

$$\bar{x} = \tfrac{1}{216}\{72{\cdot}34\times 0+86{\cdot}81\times 1+43{\cdot}40\times 2+11{\cdot}57\times 3+1{\cdot}74\times 4+0{\cdot}14\times 5\}$$
$$= 1.$$

(ii) From table 6 C, the mean number of heads per toss,

$$\bar{x} = \tfrac{1}{832}\{26\times 0+130\times 1+260\times 2+260\times 3+130\times 4+26\times 5\}$$
$$= 2\tfrac{1}{2}.$$

(iii) From table 6 D it is only possible to obtain an approximate value for the mean number of black balls per sample because of the combined probability for '3 or more'. This approximate value

$$\bar{x} = \tfrac{1}{100}\{59{\cdot}8\times 0+31{\cdot}5\times 1+7{\cdot}5\times 2+1{\cdot}2\times 3\}$$
$$= 0{\cdot}501.$$

The above three values (i), (ii) and (iii) of $\bar{x}$ are examples of the following valuable property of the Binomial distribution which will be proved formally in § 86.

If a variate x takes the values 0, 1, 2, ..., n with probabilities equal respectively to the terms, arranged in the usual way, of the binomial expansion of $(q+p)^n$, where $q+p = 1$, then the mean value of x is np.

Thus (i) in table 6 B, $n = 6$, $p = \frac{1}{6}$ and the mean $np = 1$,

(ii) in table 6 C, $n = 5$, $p = \frac{1}{2}$ and the mean $np = 2\frac{1}{2}$,

and (iii) in table 6 D, $n = 10$, $p = 0{\cdot}05$ and the mean $np = 0{\cdot}5$.

In chapter 7 frequent use is made of the facts that

the mean $\bar{x}$ or $\mu = np$,

and the standard deviation $\sigma = \sqrt{(npq)}$. (See § 155).

86. The mean and variance of the Binomial distribution. By setting down the Binomial distribution as shown in table 6 H its mean and variance can easily be deduced.

TABLE 6H

(i) Variate x	(ii) Probability or relative frequency f	(iii) fx (i)×(ii)	(iv) fx^2 (i)×(iii)
0	q^n	0	0
1	$nq^{n-1}p$	$nq^{n-1}p$	$nq^{n-1}p$
2	$\frac{n(n-1)}{1.2}q^{n-2}p^2$	$\frac{n(n-1)}{1}q^{n-2}p^2$	$\frac{2n(n-1)}{1}q^{n-2}p^2$
3	$\frac{n(n-1)(n-2)}{1.2.3}q^{n-3}p^3$	$\frac{n(n-1)(n-2)}{1.2}q^{n-3}p^3$	$\frac{3n(n-1)(n-2)}{1.2}q^{n-3}p^3$
⋮	⋮	⋮	⋮
n	p^n	np^n	n^2p^2
Total	Σf	Σfx	Σfx^2

Now $\Sigma f = 1$,

$$\Sigma fx = np\left\{q^{n-1}+\frac{(n-1)}{1}q^{n-2}p+\frac{(n-1)(n-2)}{1.2}q^{n-3}p^2+\ldots+p^{n-1}\right\}$$

$$= np(q+p)^{n-1}$$

$$= np,$$

and
$$\Sigma fx^2 = np\left\{q^{n-1}+\frac{2(n-1)}{1}q^{n-2}p+\frac{3(n-1)(n-2)}{1.2}q^{n-3}p^2+\ldots+np^{n-1}\right\}$$

$$= np\left\{q^{n-1}+\frac{(n-1)}{1}q^{n-2}p+\frac{(n-1)(n-2)}{1.1}q^{n-3}p^2+\ldots+p^{n-1}+\frac{(n-1)}{1}q^{n-2}p+\frac{2(n-1)(n-2)}{1.2}q^{n-3}p^2+\ldots+(n-1)p^{n-1}\right\}$$

$$= np\{(q+p)^{n-1}+(n-1)p(q+p)^{n-2}\}$$

$$= np\{1+(n-1)p\}$$

$$= np(1-p)+(np)^2.$$

Thus the mean, $$\frac{\Sigma fx}{\Sigma f} = np$$

and the variance, $$\frac{\Sigma fx^2}{\Sigma f} - \left(\frac{\Sigma fx}{\Sigma f}\right)^2 = npq.$$

This implies that the standard deviation $\sigma = \sqrt{(npq)}$.

87. How to estimate the proportion of black balls in a sampling-bottle. Let us now suppose that the proportion of black balls in a sampling-bottle is unknown and that 100 samples of 8 drawn from the bottle give the following distribution:

No. of black balls in the samples of 8	x	0	1	2	3	4	5 or more	Total
No. of samples	f	32	41	20	6	1	0	100

The mean number of black balls per sample

$$\bar{x} = \tfrac{1}{100}\{32 \times 0 + 41 \times 1 + 20 \times 2 + 6 \times 3 + 1 \times 4\}$$
$$= 1{\cdot}03,$$

and in this case $n = 8$.

Hence $\bar{x} = np$ gives $1{\cdot}03 = 8p$ and thus

$$p = 0{\cdot}13 \text{ approximately.}$$

This means that 13% of the balls in the bottle are black. Here then is a valuable method of estimating the proportion p of defective items in a bulk by examining random samples of n items.

88. Estimation of the proportion of defective items in a bulk. *The items produced by a certain machine were checked by examining samples of 12. The following table shows the distribution of 50 samples according to the number of defective items they contained:*

No. of defectives in a sample of 12	0	1	2	3	4	5 or more	Total
No. of samples	18	19	9	3	1	0	50

Calculate the mean number of defectives per sample and, assuming that the binomial law applies, estimate the proportion of defective items in the whole bulk produced by the machine.

By an easy calculation the mean number of defectives per sample is 1, and since $n = 12$, it follows that $p = \frac{1}{12}$.

Thus $100p = 8\frac{1}{3}\%$ is the estimated proportion of defective items in the whole bulk.

89. Exercises.

1. Razor blades of a certain kind are sold in packets of five. The following table shows the frequency distribution of 100 packets according to the number of faulty blades contained in them:

No. of faulty blades	0	1	2	3	4	5
No. of packets	80	17	2	1	0	0

Calculate the mean number of faulty blades per packet and, assuming that the binomial law applies, estimate the probability that a blade taken at random from any packet will be faulty. (J.M.B.)

2. The production of an electrical component is checked by examining samples of 6. The following table shows the frequency distribution of 40 samples according to the number of defective components contained in each:

No. of defective components in sample of 6	0	1	2	3	4	5 or 6
No. of samples	12	16	8	3	1	0

Estimate the proportion of defective components in the whole output.

7

The Normal distribution

90. From Binomial to Normal distribution. Before studying the Normal distribution in detail let us first examine diagrams which illustrate its connection with the Binomial distribution.

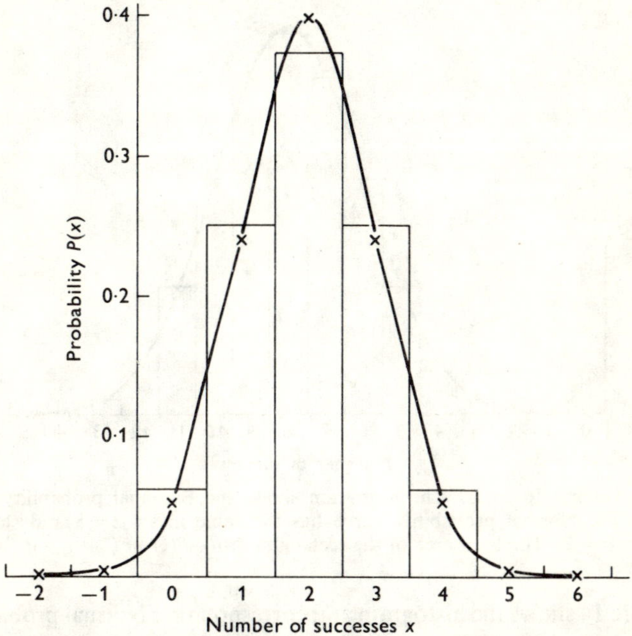

Fig. 13. Tossing 4 coins (see § 79). For each the probability of success $p = \frac{1}{2}$. The histogram shows the Binomial probability distribution. The Normal probability curve has the same mean $\mu = 2$ and standard deviation $\sigma = 1$. The total area of the rectangles $P(0)+P(1)+P(2)+P(3)+P(4) = 1$.

The histogram of fig. 13 shows the Binomial probability distribution of the number of successes (heads) when 4 coins are tossed together. In this case $p = q = \frac{1}{2}$ and the histogram is symmetrical about the mean $\mu = 2$ with standard deviation $\sigma = 1$. In fig. 13 also, the *Normal probability curve* with mean $\mu = 2$ and standard deviation $\sigma = 1$ is shown. One notices immediately that the area under the curve between $1\frac{1}{2}$ and $2\frac{1}{2}$ is approximately equal to the area of the rectangle which represents $P(2)$. Also the area under the curve between $\frac{1}{2}$ and $1\frac{1}{2}$ is approximately equal to $P(1)$ and the area under the curve to the left of $\frac{1}{2}$

is approximately equal to $P(0)$. Fig. 13, therefore, illustrates the important idea that the *Normal probability function* can be used to obtain approximations for Binomial probabilities provided we remember that the latter only exist for the *discrete* values 0, 1, 2, ... and that their approximations are found from the *continuous* function by the respective areas for the ranges 'to the left of $\frac{1}{2}$', '$\frac{1}{2}$ to $1\frac{1}{2}$', '$1\frac{1}{2}$ to $2\frac{1}{2}$', ...

This introduction of $\frac{1}{2}$ is known as the *correction for continuity*.

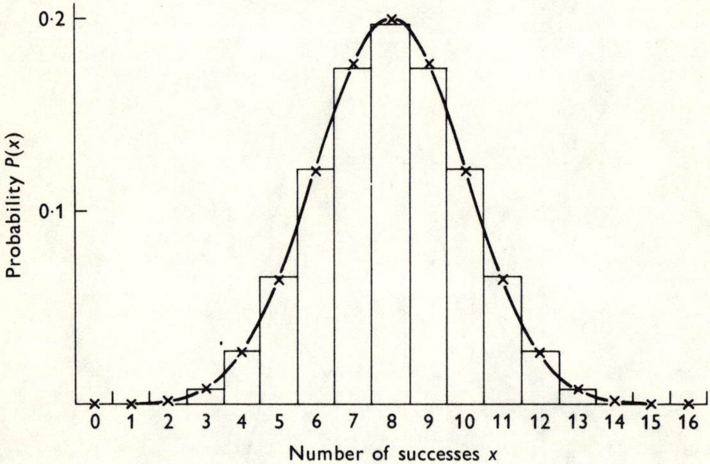

Fig. 14. Tossing 16 coins. The histogram shows the Binomial probability distribution. The Normal probability curve has the same mean $\mu = 8$ and standard deviation $\sigma = 2$. The total area of the rectangles $P(0)+P(1)+P(2)+\ldots+P(16) = 1$.

Figure 14 shows the histogram and corresponding Normal probability curve for the case of 16 coins tossed together. The greater the number of coins the nearer the histogram gets to the Normal probability curve. In general we say that, as the number of trials n increases, the Binomial distribution *tends* to the Normal.

The last statement is true even when p and q are not equal. The histogram of fig. 15 shows the Binomial probability distribution of the number of successes when the results of 9 football matches are forecast. We assume in this case that for each match win, lose or draw are equally probable. This means that $p = \frac{1}{3}$, $q = \frac{2}{3}$ and the histogram is not symmetrical. However, the Normal probability curve is close to the histogram in spite of the fact that n is not really very large. The minimum size of n for which the Normal distribution is a close approximation to the Binomial distribution depends on the value of p. The

Normal approximation to the Binomial is quite good provided *np* and *nq* are both greater than 5.

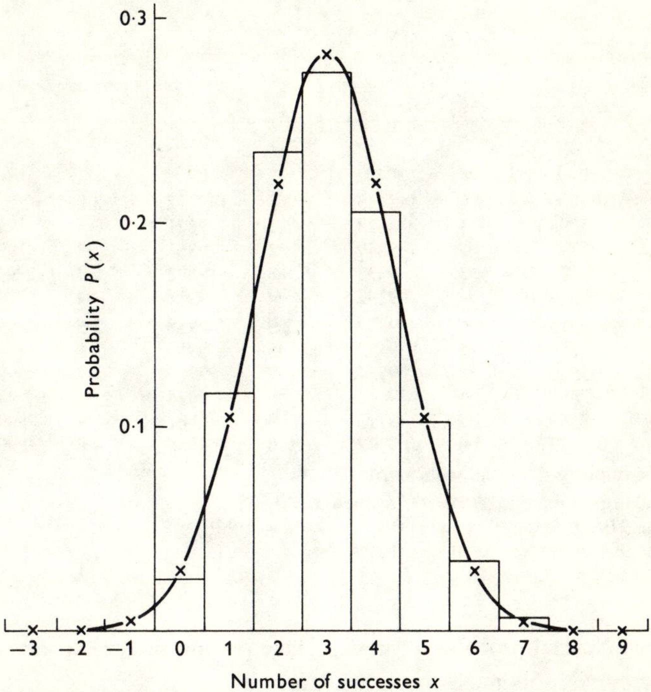

Fig. 15. Forecasting the results of 9 football matches. For each the probability of success $p = \frac{1}{3}$ (we assume). The histogram shows the Binomial probability distribution. The normal probability curve has the same mean and standard deviation.

TABLE 7A

Normal probability curve

Values of the ordinate $y = \frac{1}{\sqrt{(2\pi)}} e^{-\frac{1}{2}x^2}$ *of the*

x	*y*	*x*	*y*	*x*	*y*	*x*	*y*
0·0	0·3989	**1·0**	0·2420	**2·0**	0·0540	**3·0**	0·0044
·1	·3970	**·1**	·2179	**·1**	·0440	**·1**	·0033
·2	·3910	**·2**	·1942	**·2**	·0355	**·2**	·0024
·3	·3814	**·3**	·1714	**·3**	·0283	**·3**	·0017
·4	·3683	**·4**	·1497	**·4**	·0224	**·4**	·0012
0·5	0·3521	**1·5**	0·1295	**2·5**	0·0175	**3·5**	0·0009
·6	·3332	**·6**	·1109	**·6**	·0136	**·6**	·0006
·7	·3123	**·7**	·0940	**·7**	·0104	**·7**	·0004
·8	·2897	**·8**	·0790	**·8**	·0079	**·8**	·0003
·9	·2661	**·9**	·0656	**·9**	·0060	**·9**	·0002
1·0	0·2420	**2·0**	0·0540	**3·0**	0·0044	**4·0**	0·0001

TABLE 7B

The area A(x) to the left of a given value of x for the curve

$$y = \frac{1}{\sqrt{(2\pi)}} e^{-\frac{1}{2}x^2}$$

$$A(x) = \frac{1}{\sqrt{(2\pi)}} \int_{-\infty}^{x} e^{-\frac{1}{2}t^2}\, dt$$

x	$A(x)$	x	$A(x)$	x	$A(x)$	x	$A(x)$
0·0	0·5000	**1·0**	0·8413	**2·0**	0·9772	**3·0**	0·99865
·1	·5398	**·1**	·8643	**·1**	·9821	**·1**	·99903
·2	·5793	**·2**	·8849	**·2**	·9861	**·2**	·99931
·3	·6179	**·3**	·9032	**·3**	·9893	**·3**	·99952
·4	·6554	**·4**	·9192	**·4**	·9918	**·4**	·99966
0·5	0·6915	**1·5**	0·9332	**2·5**	0·9938	**3·5**	0·99977
·6	·7257	**·6**	·9452	**·6**	·9953	**·6**	·99984
·7	·7580	**·7**	·9554	**·7**	·9965	**·7**	·99989
·8	·7881	**·8**	·9641	**·8**	·9974	**·8**	·99993
·9	·8159	**·9**	·9713	**·9**	·9981	**·9**	·99995
1·0	0·8413	**2·0**	0·9772	**3·0**	0·9986	**4·0**	0·99997

Some important special values are:

the upper quartile $A(x) = 0·75$ when $x = 0·675$;
the 97·5th percentile $A(x) = 0·975$ when $x = 1·96$;
the 99·9th percentile $A(x) = 0·999$ when $x = 3·09$.

91. The Normal probability curve. The equation of the Normal probability curve in its most convenient form is

$$y = \frac{1}{\sqrt{(2\pi)}} e^{-\frac{1}{2}x^2}.$$

It was originally derived by Gauss as the law of the probable distribution of errors of measurement and hence it is often called the *Gaussian* or *error* curve. It is a continuous function for which x may range from $-\infty$ to $+\infty$, but the part of the curve which has real practical value lies between $x = -4$ and $x = +4$. Ordinary books of mathematical tables contain values of e^x and e^{-x} and the student will have no difficulty in using them to obtain values of the ordinate y of the Normal probability curve for given values of x. The normal probability curve, however, is of such fundamental importance in statistical analysis that it is useful to have values of y, tabulated for given values of x as in table 7A.

92. Negative values of *x*. Since y is a function of x^2, negative values of x such as -1, -2, -3 give the same positive value to y as the positive values 1, 2, 3. Thus the graph of the function is symmetrical about the y-axis.

93. The graph of the function. The graph of

$$y = \frac{1}{\sqrt{(2\pi)}}\, e^{-\frac{1}{2}x^2}$$

between $x = -4$ and $x = +4$ is shown in fig. 16. The student can construct it for himself by using the values given in table 7 A.

94. The area under the curve. The dotted rectangles shown in fig. 16 are each of unit width and their heights are the ordinates at $-3{\cdot}5$, $-2{\cdot}5$, $-1{\cdot}5$, $-0{\cdot}5$, $+0{\cdot}5$, $+1{\cdot}5$, $+2{\cdot}5$ and $+3{\cdot}5$. Their areas are, therefore, 0·0009, 0·0175, 0·1295, 0·3521, 0·3521, 0·1295, 0·0175, 0·0009. Thus the area under the graph between $x = -4$ and $x = +4$ is approximately unity. Note that the two rectangles between $x = -4$ and $x = -3$ and between $x = 3$ and 4 are so small that they have been left to the imagination.

Actually the area between $-\infty$ and $+\infty$ can be proved to be exactly unity. Mathematically this is stated as

$$\frac{1}{\sqrt{(2\pi)}}\int_{-\infty}^{\infty} e^{-\frac{1}{2}x^2}\, dx = 1.$$

The above approximation of unity for the area between -4 and $+4$ is a basis for the statement in § 91 that the part of the curve which is of real practical value is that between -4 and $+4$.

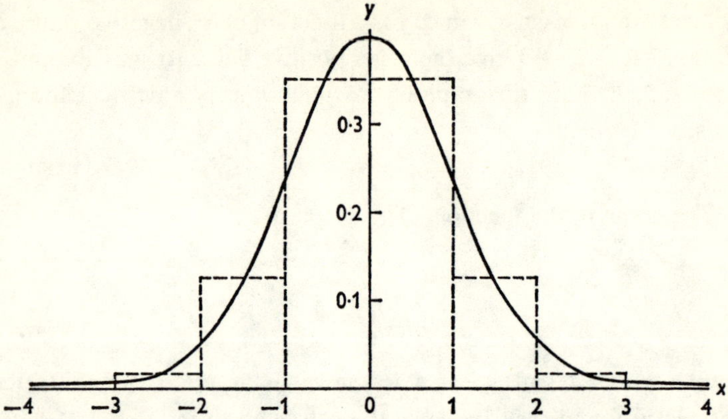

Fig. 16. The Normal probability curve. The continuous bell-shaped curve is the graph of

$$y = \frac{1}{\sqrt{(2\pi)}} e^{-\frac{1}{2}x^2}.$$

The dotted rectangles are used to show that the area under the curve is approximately unity. The rectangle between $x = 3$ and $x = 4$ is left to the imagination and similarly that between $x = -4$ and $x = -3$.

95. Exercises using Simpson's rule. Simpson's rule, that an approximation for the area A under $y = f(x)$ between two ordinates y_1 and y_3, at a distance $2h$ apart is

$$A = \tfrac{1}{3}h\{y_1 + 4y_2 + y_3\},$$

where y_2 is the mid-ordinate can be used, in conjunction with table 7 A, to obtain approximations for the area under the probability curve between given values of x. Thus, for the area between $x = 0$ and $x = 1$,

$$2h = 1, \quad y_1 = 0{\cdot}3989, \quad y_2 = 0{\cdot}3521, \quad y_3 = 0{\cdot}2420,$$

and an approximation for the area is

$$\begin{aligned} A &= \tfrac{1}{6}\{0{\cdot}3989 + 1{\cdot}4084 + 0{\cdot}2420\} \\ &= 0{\cdot}3416. \end{aligned}$$

The student will find it a useful exercise to verify, by similar calculations, that

(i) the area between $x = 1$ and $x = 2$ is approximately 0·1357,

(ii) the area between $x = 2$ and $x = 3$ is approximately 0·0214,

(iii) the area between $x = 3$ and $x = 4$ is approximately 0·00135.

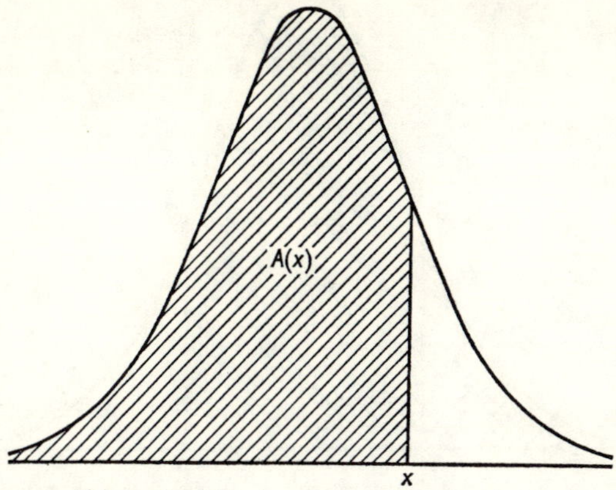

Fig. 17. The area $A(x)$ to the left of a given value of x is tabulated in table 7 B. Mathematically expressed,

$$A(x) = \frac{1}{\sqrt{(2\pi)}} \int_{-\infty}^{x} e^{-\frac{1}{2}t^2}\, dt.$$

96. The area under the graph to the left of a given ordinate. The area under the probability curve is of great practical importance. Table 7 B, shows the area $A(x)$ under the graph to the left of a given value of x as illustrated in fig. 17. Thus $A(0) = 0{\cdot}5000$ indicates that half of the area under the graph is to the left of $x = 0$ and $A(1) = 0{\cdot}8413$ indicates that 0·8413 of the area (i.e. 84·13%) is to the left of $x = 1$. Moreover, $A(1) - A(0) = 0{\cdot}3413$ indicates that 0·3413 of the area (i.e. 34·13%) is between the ordinates at $x = 0$ and $x = 1$.

97. *A*(*x*) for negative value of *x*. Table 7 B indicates that $A\ (1{\cdot}96) = 0{\cdot}975$. Thus 97·5% of the area is to the left of $x = 1{\cdot}96$ and 2·5% to the right of $x = 1{\cdot}96$. As the graph is symmetrical about the y-axis it follows that 2·5% of the area lies to the left of $x = -1{\cdot}96$ and hence

$$A(-1{\cdot}96) = 1 - A(1{\cdot}96).$$

In general it may be stated that the area to the left of a negative value of x is obtained by subtracting from unity the area to the left of the corresponding positive value of x. In symbols this is written

$$A(-x) \doteq 1 - A(x)$$

and it is illustrated diagrammatically in fig. 18.

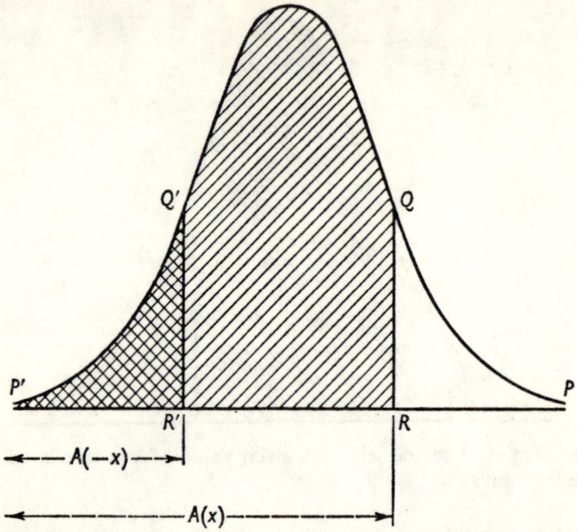

Fig. 18. Illustrates that $A(-x) = 1 - A(x)$ because area PQR = area $P'\ Q'\ R'$.

98. The mean and the standard deviation. The Normal probability distribution given in tables 7 A and 7 B has its mean situated at the origin, $x = 0$, and its standard deviation is unity. If the heights of a group of men are known to be Normally distributed about a mean of 168 cm with a standard deviation of 5 cm then heights in cm of 168, 168+5, 168+10, 168+15, 168+20 must be regarded as replacing $x = 0, 1, 2, 3, 4$ and

$$A(1) - A(0) = 0{\cdot}3413$$

indicates that 0·3413 of the total population of men of the group will probably be between 168 and 173 cm in height while

$$A(2) - A(1) = 0{\cdot}1359$$

indicates that 0·1359 of the total number will probably be between 173 and 178 cm in height and so on.

Thus from the mean 168 cm and the standard deviation 5 cm, by table 7 B, the following relative frequency distribution can be obtained:

Heights of men (cm)	Relative frequency
under 153	0·0014
153–158	0·0214
158–163	0·1359
163–168	0·3413
168–173	0·3413
173–178	0·1359
178–183	0·0214
over 183	0·0014
Total	1·0000

The histogram for the above relative frequency distribution will, of course, be exactly like fig. 16. It is important to realize that if any man is taken at random from the above group the probability that his height is between 168 and 173 cm is 0·3413 (approximately $\frac{1}{3}$) and the probability that his height is above 183 cm is 0·0014 (approximately $\frac{1}{100}$).

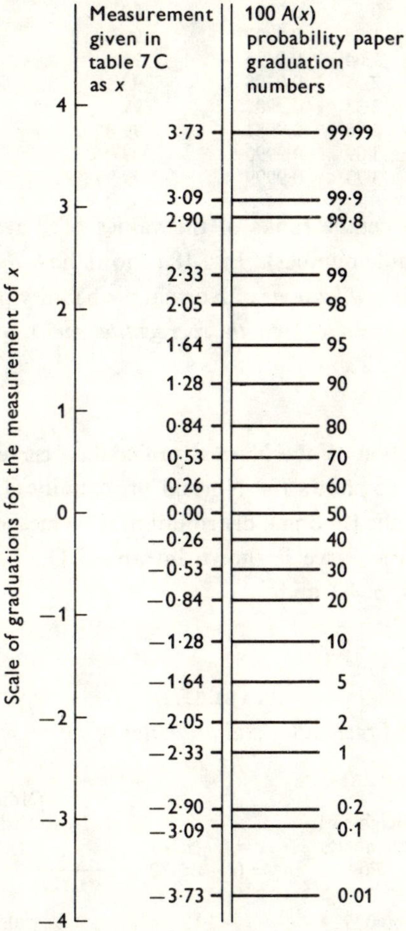

Fig. 18*a*. Shows the way in which the graduations of Normal probability paper are measured and labelled.

98a. The construction of Normal probability paper. Table 7 C is a modified form of table 7 B in which some specially chosen values of x are given to 2 decimal places. Note that the values of x give the values 0·5000, 0·6026, 0·7019, ... for $A(x)$.

TABLE 7C

The construction of Normal probability paper

x	$A(x)$	Percentile rank of x 100 $A(x)$
0·00	0·5000	50
0·26	0·6026	60
0·53	0·7019	70
0·84	0·7995	80
1·28	0·8997	90
1·64	0·9495	95
2·05	0·9798	98
2·33	0·9901	99
2·90	0·9981	99·8
3·09	0·9990	99·9
3·73	0·9999	99·99

and hence the percentile ranks of the values of x are 50, 60, 70, ... (to the nearest whole number). Fig. 18*a* shows how *the values of x are used to measure the graduations* of Normal probability paper. When the graduations have been drawn *the percentile ranks of x are used to label them.*

98b. The construction of the Normal probability curve for a Binomial distribution. Fig. 15 shows the Normal probability curve which is an approximation to the Binomial distribution. The method of calculating the ordinates of the curve is shown in table 7 D. For the Binomial distribution $n = 9$, $p = \frac{1}{3}$ and

TABLE 7D

Heights of rectangles and ordinates of curve shown in fig 15

Number of successes r	Heights of rectangles $P(r)$	$x = (r-\mu)/\sigma$ $= (r-3)/\sqrt{2}$	Ordinates of Normal distribution $\frac{1}{\sigma\sqrt{(2\pi)}} e^{-\frac{1}{2}x^2} = \frac{1}{\sigma} A(x)$
0	0·026	−2·12	0·030
1	0·117	−1·41	0·104
2	0·234	−0·71	0·219
3	0·273	0·00	0·282
4	0·205	0·71	0·219
5	0·102	1·41	0·104
6	0·034	2·12	0·030
7	0·007	2·83	0·005
8	0·001	3·54	0·001
9	0·001	4·24	0·000

hence the mean $\mu = 3$ and the standard deviation $\sigma = \sqrt{2}$. The ordinates of the Normal probability curve are given by

$$y = \frac{1}{\sigma\sqrt{(2\pi)}} e^{-\frac{1}{2}(r-\mu)^2/\sigma^2}$$

where r takes the values 0, 1, 2, ..., 9. Thus, if $x = (r-\mu)/\sigma$, y can be obtained from table 7 B by multiplying $A(x)$ by $1/\sigma$.

99. Exercises on the use of table 7B.

1. A student who cycles regularly from A to B states that the times of the journey are Normally distributed about a mean of 60 min with standard deviation 5 min. What is the probability that, on any occasion, the journey will take (i) more than 70 min? (ii) less than 55 min?

2. At a sugar refinery a machine packs automatically 1 kg cartons of sugar. When the machine is in operation the cartons emerge from it in a continuous stream ready for sale. In order to check that the machine is giving correct weight, sample cartons are taken at random from it and weighed accurately. From these random samples it is estimated that the mean weight of carton delivered by the machine is 1·02 kg and that the standard deviation of the weights of the cartons is 0·01 kg. Assuming that the weights are Normally distributed, estimate the percentage of cartons the machine delivers underweight.

3. The heights in mm of a certain type of compression spring produced in a workshop may be assumed to be Normally distributed about a mean 100·5 mm with standard deviation 0·2 mm. Springs which are less than 100 mm are useless and have to be discarded. Estimate

(i) the percentage of springs discarded;

(ii) the central 50% of production (i.e. the interquartile range).

4. The mean *length of life* of a certain type of television tube is 1600 h with a standard deviation of 250 h. Assuming that the lengths of life of television tubes of this type are Normally distributed, copy, and complete, the following frequency table and use it to draw a histogram:

Length of life (h)	Less than 750	750–1000	1000–1250	1250–1500	1500–1750	1750–2000	2000–2250	More than 2250
Percentage of tubes								

5. The *recovery time* of an aircraft is the time between landing and being ready to fly again. In an investigation on a certain type of aircraft it was found that the recovery times were Normally distributed about a mean of 22 min with a standard deviation of 5 min. Make out a frequency table showing the percentage of aircraft with recovery times 5 min to 10 min, 10 min to 15 min, 15 min to 20 min, etc., and draw the histogram.

6. Tests on electric lamps of a certain type indicated that their lengths of life could be assumed to be Normally distributed about a mean 1850 h with standard deviation 100 h. Estimate the percentage of lamps which can be expected to burn (i) more than 2000 h, (ii) less than 1500 h.

7. Steel rods which are being manufactured to a specification of 30 mm diameter are acceptable if they are within the *tolerance limits* 30·1 mm and 29·9 mm. If the diameters can be assumed to be Normally distributed about a mean 30·02 mm with standard deviation 0·05 mm, estimate the percentage of rods that will be rejected (i) oversize, (ii) undersize.

8. Limit gauges are used to reject all components in which a certain dimension is greater than 60·4 mm or less than 59·6 mm. It is found that $2\frac{1}{2}\%$ are rejected oversize and $2\frac{1}{2}\%$ are rejected undersize. Assuming the dimensions are Normally distributed find the mean and the standard deviation correct to 2 decimal places.

Use your mean and standard deviation to estimate the percentage of rejects if the limits are changed to (i) 60·3 mm to 59·7 mm, (ii) 60·5 mm to 59·5 mm.

9. A machine makes electrical resistors having a mean resistance of 100 ohms with standard deviation of 5 ohms. It is observed that a certain firm rejects approximately 11% of the resistors ($5\frac{1}{2}\%$ over and $5\frac{1}{2}\%$ under) because they are not within its *tolerance limits*. Assuming the distribution of values to be Normal, estimate the tolerance limits employed by the firm.

8

Regression

100. A bivariate distribution. Table 8 A shows, side by side, the number of vehicles, x, on the roads of Great Britain and the total casualties, y, in road accidents for the years 1945–1954.

TABLE 8A

Year	Vehicles (millions) x	Casualties (thousands) y
1945	2·6	138
1946	3·1	163
1947	3·5	166
1948	3·7	153
1949	4·1	177
1950	4·4	201
1951	4·6	216
1952	4·9	208
1953	5·3	226
1954	5·8	238

The number of vehicles is seen to have steadily increased, and the number of road accidents has also increased though not quite so regularly. The ten (x, y) pairs of values are an example of a *bivariate distribution.* The data is continued for 1955–67 in § 116, Ex. 16.

101. The scatter diagram: direct correlation. If the value y is plotted on a graph against its corresponding value x as shown in fig. 19, a *scatter diagram* is obtained.

Although the points on the scatter diagram do not fall exactly along a straight line they fall within quite a narrow belt. Small values of y correspond to small values of x, large values of y to large values of x, and x and y are said to be *directly correlated.* The exercises at the end of this chapter will provide the student with further examples of *direct correlation.*

102. Inverse correlation. In the bivariate distribution of x and y shown in table 8 B, *small* values of y correspond to large values of x and vice versa. In the three years following the opening of a new television transmitter to serve a certain area, the cinema admissions *decreased* as the number of television sets owned by the people *increased.* This is known as *inverse correlation.* In the scatter diagram (fig. 20), the points

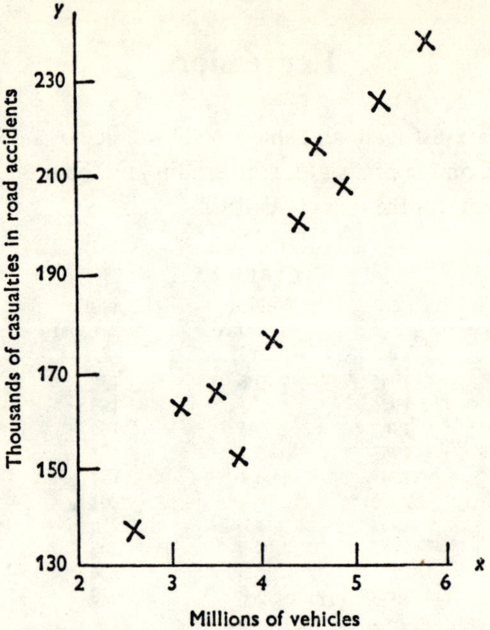

Fig. 19. Scatter diagram showing direct correlation.

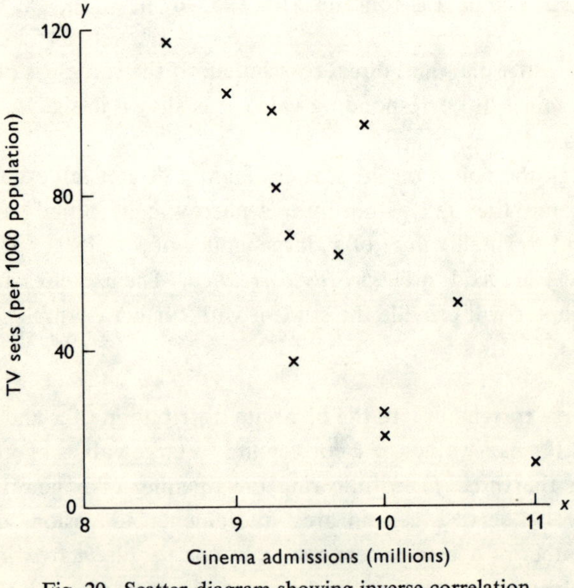

Fig. 20. Scatter diagram showing inverse correlation.

lie within a fairly well-defined belt which is downward sloping for increasing values of x.

TABLE 8B

Cinema admissions and television sets

Year and quarter	Towns served by a new TV transmitter: Cinema admissions (millions) x	TV sets (per 1000 population) y	Towns not served by the new TV transmitter: Cinema admissions (millions) z	TV licences (per 1000 population)
1950 1	11·0	12	12·1	—
2	10·0	18	11·2	—
3	10·0	24	11·6	—
4	9·4	37	10·6	—
1951 1	10·5	52	12·1	—
2	9·7	64	11·8	—
3	9·4	69	11·5	—
4	9·3	81	10·9	1
1952 1	9·9	98	11·6	2
2	9·3	101	11·1	3
3	9·0	106	11·3	3
4	8·6	119	10·5	4

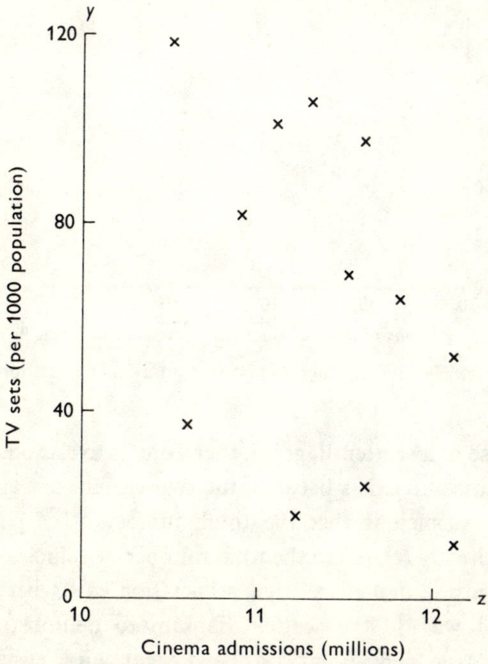

Fig. 21. Scatter diagram of a bivariate distribution in which the variables are not correlated.

103. Absence of correlation and spurious correlation. Fig. 21 is the scatter diagram of the bivariate distribution of y and z of the last table. One would not expect correlation in this case because it is unlikely that the *cinema admissions* in one group of towns should be related to the *number of TV sets* in a completely different group of towns. This is confirmed by the scatter diagram, the points of which do not lie within any well-defined belt.

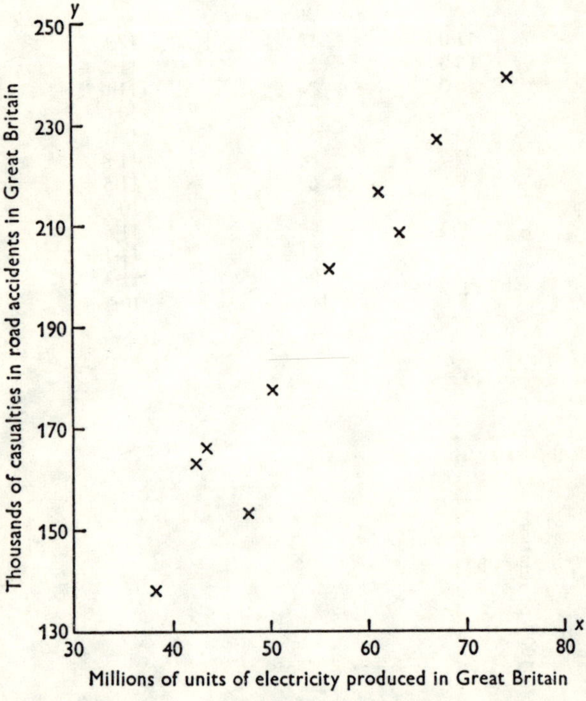

Fig. 22. Scatter diagram showing spurious correlation.

The purpose of a scatter diagram, therefore, is to establish whether or not any relationship exists between the two variables x and y. Fig. 19 confirms our suspicions that the total number of casualties in road accidents is directly related to the total number of vehicles on the roads. We assume, in the first place, that association exists between the two variables and we use the scatter diagram to demonstrate that our assumption is true. This point is stressed because it is very easy to draw scatter diagrams which are sheer nonsense. For the years 1945–54 the

Total Electricity Production of Great Britain in millions of units as follows:

38·6, 42·7, 43·9, 48·0, 50·6, 56·5,
61·5, 63·9, 67·4, 74·7.

Fig. 22 shows these values plotted as x against the total casualties in road accidents plotted as y. Although the points in this scatter diagram

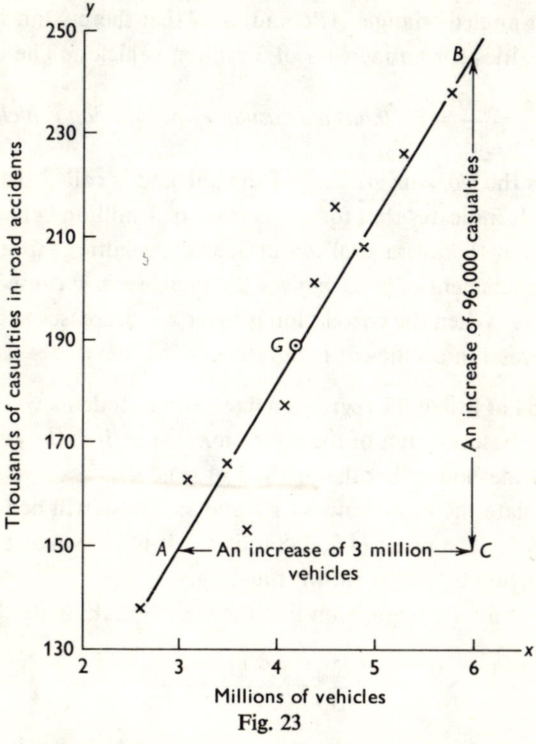

Fig. 23

lie within a well-defined belt it is ridiculous to conclude that the road casualties are related in any way to the total electricity generated. Fig. 22 is an example of *spurious correlation.* If x and y continuously increase or decrease over a period of time even though they are obviously unrelated in any way they may appear in a scatter diagram to be correlated.

Finally another important point about correlation must be stressed. If association is proved to exist between two factors it must not be assumed that variation in one *causes* variation in the other. The two factors may be related because they both depend on a third factor. In exercise 1, § 110, the two factors *ability at French* and *ability at German* both depend on the third factor *linguistic ability*.

104. The regression line and the regression coefficient. Fig. 23 shows the scatter diagram of fig. 19 with a straight line fitted through the middle of the ten points. It is called the *regression line*. It indicates how the *thousands of casualties*, y, vary with the *millions of vehicles*, x.

The right-angled triangle ABC indicates that there is an increase of 96,000 casualties for an increase of 3 million vehicles. The ratio

$$\frac{CB}{AC} = \frac{96}{3} = 32 \textit{ thousand casualties per million vehicles}$$

is known as the *slope* or *gradient* of the line and is called the *regression coefficient*. It indicates that for an increase of 1 million vehicles there is an increase on the average of about 32,000 casualties. In general, the regression coefficient of y on x gives the increase in y caused by a unit increase in x. When the correlation is inverse y decreases as x increases and the regression coefficient is negative.

105. Method of fitting the regression line. Some students will be content to judge the best position of the regression line entirely *by eye*. A more satisfactory method called the method of least squares is as follows:

(i) Calculate the mean values of x and y. These will be found to be 4·2 and 188·6. The point (4·2, 188·6) is G. It is known as the *mean of the array* of points, and is shown thus: $\odot$.

(ii) Now draw the regression line through G making its slope

$$\frac{S_{xy}}{S_x^2} = \frac{29{\cdot}11}{0{\cdot}898} = 32$$

where S_{xy} and S_x^2 are calculated by the method shown in table 9 A, § 112.

106. The two regression lines for more widely scattered points. In most scatter diagrams the points are so widely dispersed that two regression lines are possible, one for estimating y for given values of x known as *the line of regression of y on x*, the other for estimating x for given values of y known as *the line of regression of x on y*. Fig. 24 is a reproduction of fig. 20 with the two regression lines fitted. *AB is the line of regression of x on y*, while CD is the line of regression of y *on* x. AB is used to estimate the cinema admissions for a given number of TV sets per

1000 population. *CD* would be used to estimate the TV sets per 1000 population for a given number of cinema admissions. As the effect of the new form of entertainment on the old is of greater interest than the effect of the old on the new, *AB* is of greater use than *CD*.

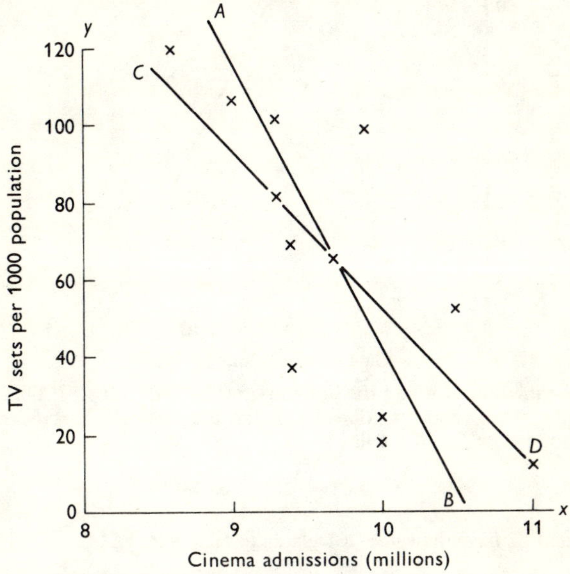

Fig. 24. *AB* is the line of regression of *x* on *y*. *CD* is the line of regression of *y* on *x*. In this example *AB* is of greater interest than *CD*.

107. To draw the line of regression of *x* on *y*.

(i) Calculate the mean of the array (9·68, 65·08) and plot it thus: $\odot$.

(ii) Now draw the regression line through the mean of the array in such a position that the *sum of the squares of the residuals* is a *minimum.* This is explained in fig. 25.

The line of regression of *x* on *y* shows how *x* depends on *y*. In this case *y* is said to be the *independent variable* while *x* is the dependent variable. The method of fitting the line is based on the assumption that the values of *y* are known accurately and that any errors that occur are in the values of *x*. The residuals, in fact, represent the errors in the values of *x*.

108. To draw the line of regression of *y* on *x*. The line of regression of *y* on *x* is drawn through the mean of the array, as in fig. 26, with the equation

$$y-65{\cdot}08 = \frac{S_{xy}}{S_x^2}(x-9{\cdot}68).$$

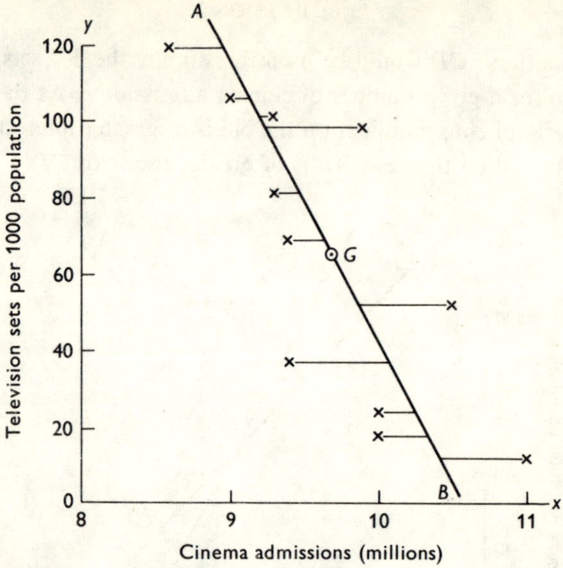

Fig. 25. The method of fitting the line of regression of x on y. If lines are drawn parallel to the x-axis from the points to the line they are called *residuals*. It can be proved that the equation of the line is

$$x - 9{\cdot}68 = \frac{S_{xy}}{S_y^2}(y - 65{\cdot}08),$$

where S_{xy} and S_y have the values calculated in table 9 B, § 113.

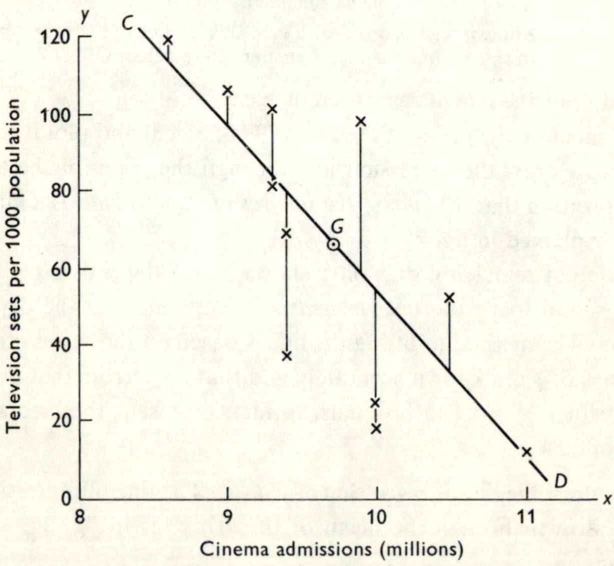

Fig. 26. The method of fitting the line of regression of y on x. The residuals in this case are parallel to the y-axis and the line is drawn through G making the *sum of the squares of the residuals* a *minimum*. The equation of the line is given in § 108.

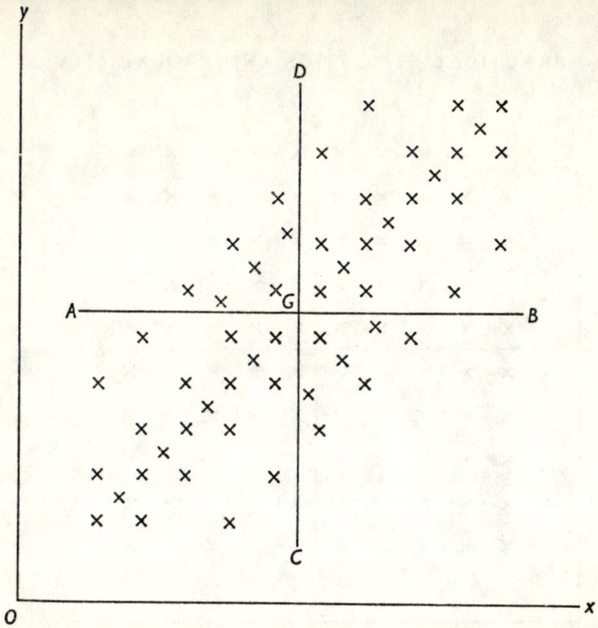

Fig. 27. Direct (or positive) correlation indicated by dichotomy. The lines *AB* and *CD* are drawn through the mean of the array, *G*, parallel respectively to the axes *x* and *y*. In this diagram a *high proportion* of the points lie in the quadrants *BGD* and *AGC*. This is an indication that *x* and *y* are directly correlated.

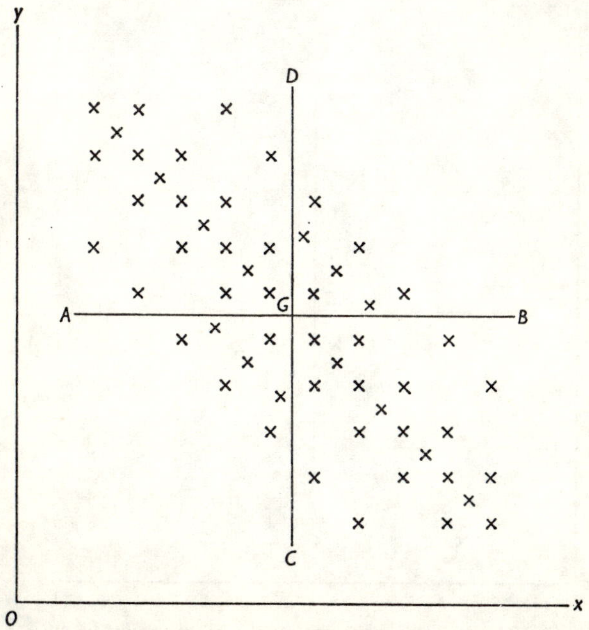

Fig. 28. Inverse (or negative) correlation indicated by dichotomy. In this case only a *small proportion* of the points lie in the quadrants *BGD* and *AGC*. This indicates that *x* and *y* are inversely correlated.

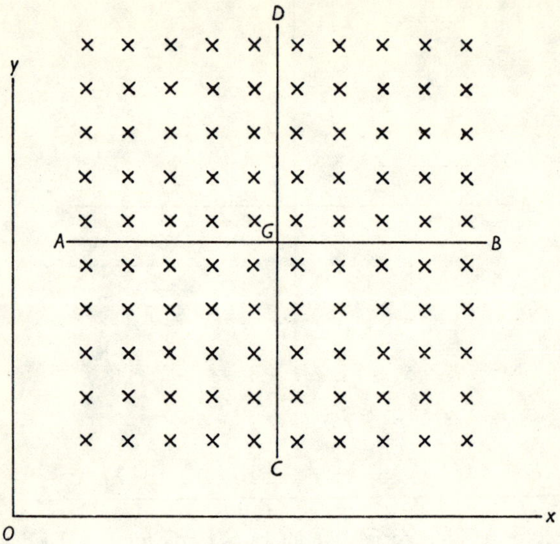

Fig. 29. Complete absence of correlation. Equal proportions of the points in all four quadrants.

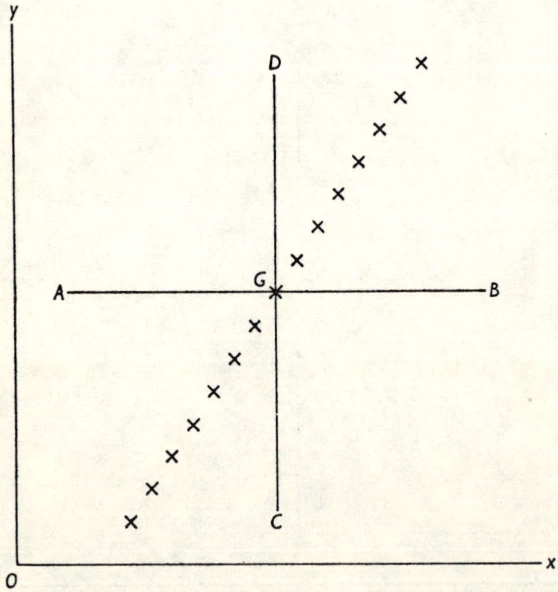

Fig. 30. Perfect direct correlation. All the points lie on a straight line.

109. Dichotomy. If the two variables of a bivariate distribution are very closely correlated the points of the scatter diagram will lie within a very narrow belt and the angle between the two regression lines will be very small. For more widely dispersed points the angle between the regression lines is greater and the correlation is said to be less marked.

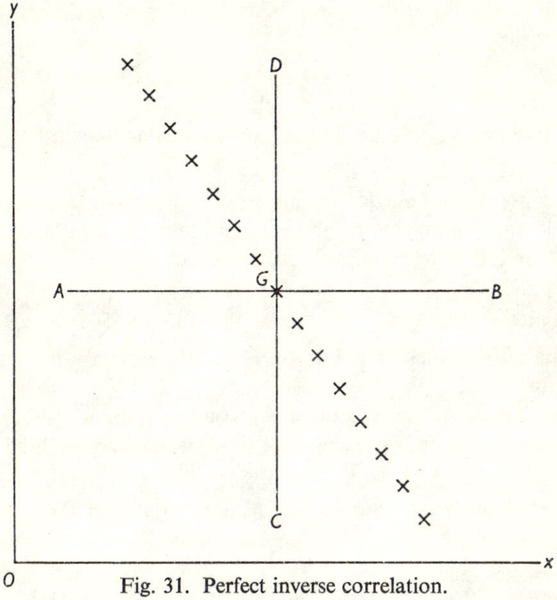

Fig. 31. Perfect inverse correlation.

An alternative approach to the assessment of the degree of correlation between x and y is by *dichotomy*. This is illustrated and fully explained in figs. 27–31. In these diagrams lines AB and CD are drawn through the mean of the array, G, parallel respectively to the axes of x and y. The points are thus divided into four quadrants. The values of x are *dichotomized* by CD, that is to say, divided into two classes. The values of y are dichotomized by AB and this process is called dichotomy (pronounced *die-cot-om-ee*).

110. Exercises.

The existence of correlation should not really be based on the evidence of less than thirty pairs of values. Dealing with such a large number of pairs is quite a long and tedious procedure. In the following exercises small numbers of pairs of values are given in order to illustrate the principles of the method without making it wearisome in its application.

1. The following table shows the examination marks of seven students in

French and German. Construct a scatter diagram and draw a line of best fit:

French	12	24	30	34	47	58	68
German	32	44	47	58	73	72	88

State any conclusion that might be drawn. (London.)

2. The following table gives the marks of eight students in each of two examinations in Botany and Zoology. Construct a scatter diagram and draw a line of best fit:

Botany	39	61	49	64	42	72	52	57
Zoology	44	62	54	70	46	76	60	64

State, with a reason, any (*a*) general, (*b*) particular conclusion you could reach. (London.)

3. In an experiment a vertical spring fixed at its upper end was stretched by the application of weights to its lower end and the lengths of the spring measured. The following readings were obtained:

Load (kg)	0	1	2	3	4	5	6	7	8
Length (cm)	12	12·5	13·1	13·8	14·5	15·2	15·7	16·2	16·7

Plot these pairs of values as points on a graph and draw the straight line which best fits them.

Assuming that the relation between the load and the length continues to hold for greater loads, find the length of the spring when the load is 12 kg. (London.)

4. Seven students were examined in Mathematics and Physics and their marks were:

Mathematics	38	51	19	53	39	38	66
Physics	50	72	36	64	52	56	80

Construct a scatter diagram and a line of best fit. Give with reasons any (*a*) particular, (*b*) general conclusion you could reach. (London.)

5. *Index numbers of wholesale prices* (X) *and retail prices* (Y)

	Jan.	Feb.	Mar.	Apr.	May	June
X	80	79	77	76	76	76
Y	84	82	81	82	81	86
	July	Aug.	Sept.	Oct.	Nov.	Dec.
X	74	72	70	71	69	70
Y	83	82	82	82	82	81

Plot a scatter diagram and draw the line of regression of Y on X.

Hence, state the average change in Y corresponding to unit positive change in X. (London.)

6. In ten areas, the percentage of dwellings overcrowded (X) and the infant mortality (Y) were as follows:

X	13	33	12	40	12	7	20	4	15	26
Y	124	151	124	156	128	78	127	104	127	144

Plot a scatter diagram, draw the line of regression of Y on X and obtain the coefficient of regression of Y on X. Explain its meaning. (London.)

7. The following table gives the percentage of sand in soil at different depths:

x (depth in mm)	0	150	300	450	600	750	900	1050	1200
y (% sand)	80·6	63·0	64·3	62·5	57·5	59·2	40·8	46·9	37·6

On a scatter diagram show the regression lines (labelled *a* and *b*) which you would use

(*a*) to predict the depth from the percentage sand;

(*b*) to estimate the dependence of percentage of sand on depth. (London.)

8. A general knowledge test consisting of a hundred questions was given to fifteen boys of different ages with results as follows:

Boy	Age: Years	Age: Months	No. of questions correct
A	11	7	18
B	11	1	19
C	12	8	23
D	12	0	26
E	13	5	25
F	13	6	31
G	14	9	24
H	15	3	32
I	14	7	28
J	15	6	25
K	15	9	33
L	15	7	31
M	16	11	36
N	17	1	32
O	16	10	40

Plot a scatter diagram of the numbers of question correct (y) against the age in months (x). Sketch in the line of regression of y on x.

State, with reasons, which boy deserves the prize for the best performance taking age into consideration. (J.M.B.)

9. *Percentage shrinkage in samples of cloth after washing, in directions along and across the cloth*

Along (x)	Across (y)	Along (x)	Across (y)
12	5	7	5
4	2	12	7
10	5	18	10
10	8	14	7
11	6	14	8
10	8	8	4
6	3	11	6
6	4	17	8
6	3	21	11
13	5	12	9

Plot a scatter diagram to illustrate the data in the table, and sketch in the line of regression of x on y.

A roll of cloth is sampled by cutting a narrow test strip right across the roll. The strip proves to have a percentage shrinkage of 7. Use your regression line to obtain an estimate of the percentage shrinkage to be expected along the cloth.

It is desired to cut from the roll a piece of cloth which may be expected to shrink to 10 cm square after washing. Describe how this piece should be cut. (J.M.B.)

10. Five groups of locusts, each containing 120, were exposed to a lethal spray in various concentrations. The deaths resulting were as follows:

Concentration (multiple of standard)	1·2	1·4	1·6	1·8	2·0
No. of deaths	38	52	46	76	66

For each concentration, find the percentage of locusts dying. Plot this percentage against the concentration on a scatter diagram and sketch in the line of regression, taking concentration as your independent variable. (J.M.B.)

11. The 1% sample of a National Census shows the ages of husband and wife to be related as shown below:

Age of wife	Age of husband 20–	30–	40–	50–	60–	70–
20–	12	7	—	—	—	—
30–	1	18	8	1	—	—
40–	—	2	19	6	1	—
50–	—	—	2	14	5	—
60–	—	—	—	1	8	2
70–	—	—	—	—	1	3

(1 unit = 100,000 couples)

For each range of age of husband calculate the mean age of wife. Plot your results on a regression graph and sketch in the regression line.

Estimate the mean age of wife for husband aged just 60 years.

(*Note.* The means may be calculated to the nearest year and the oldest classes may be taken as 70–80.)

12. The weight and height of each boy in a certain school were recorded and the pairs of measurements were found to be related as follows:

Height of boy (mm)	Weight of boy (kg) 25–	31–	37–	43–	49–	55–	61–	67–	73–	79–
1300–	—	1	—	—	—	—	—	—	—	—
1375–	4	12	4	—	—	—	—	—	—	—
1450–	—	22	21	3	—	—	—	—	—	—
1525–	—	6	28	14	1	1	—	—	—	—
1600–	—	1	1	24	17	4	—	—	1	—
1675–	—	—	—	4	22	18	2	3	—	—
1750–	—	—	—	—	1	14	10	4	—	—
1825–	—	—	—	—	—	2	2	1	—	—
1900–	—	—	—	—	—	—	—	—	—	—

For each range of height calculate the mean weight in kg. Plot your results on a regression graph and sketch in the regression line.

Estimate the mean weight in kg of boys of height 1750 mm. (J.M.B.)

13. The average prices, for each of the months January to June, of two commodities in Naira are given in the following table:

	Jan.	Feb.	Mar.	Apr.	May	June
A	3	4	4	2	6	7
B	7	9	12	4	22	19

Plot the pairs of values (A, B) in a scatter diagram. It has been suggested that the prices B vary linearly with the prices A. Do you agree with this conclusion? Give reasons. (W.A.E.C.)

9

Correlation by product-moments

111. The coefficient of correlation. Correlation was first investigated graphically by Sir Francis Galton. For this reason, the scatter diagram is often called the *Galton graph.* In 1896, Karl Pearson introduced a method of assessing correlation by means of a numerical *coefficient of correlation.* This is calculated by a formula based upon a mathematical study of the regression lines. It lies between +1 and −1 and is positive for direct correlation and negative for inverse correlation. Values of the coefficient near to unity such as 0·85 or 0·90 indicate a high degree of correlation. Values near to zero such as 0·15 or 0·20 indicate an absence of correlation except when the coefficient has been calculated from a large number of pairs of values of the bivariate distribution. The coefficient of correlation for the *millions of vehicles*, x and the *thousands of casualties*, y, of § 100 will be calculated in this chapter and shown to be 0·96. This indicates a marked degree of correlation between the variables and the angle between the regression lines (when determined mathematically) is very small. The coefficient of correlation for the *cinema admissions*, x, and *TV sets*, y, of § 102 will be shown to be −0·75. This indicates quite a high degree of inverse correlation. As the acute angle between the regression lines increases from 0 to 90° the numerical value of the coefficient of correlation decreases from 1 to 0. A coefficient of +1 indicates perfect direct correlation, a coefficient of −1 indicates perfect inverse correlation, and a coefficient of 0 indicates complete absence of correlation. See figs. 29–31.

112. Calculation of the coefficient of correlation. The method of calculating the coefficient of correlation is shown in table 9 A. The given values of x and y together with their totals are given in the first two columns. The next four columns are used for calculating the standard deviation, S_x, of the ten values of x, and the standard deviation, S_y, of the ten values of y. The last column contains the products of deviations, $d_x d_y$, whose mean value is called the *covariance*, S_{xy}.

113. The example of inverse correlation. Table 9 B shows the calculation of the coefficient of correlation for the *cinema admissions*, x, and the

TABLE 9A

Calculation of coefficient of correlation (direct)

Millions of vehicles x	Thousands of casualties y	Deviations of x from the mean 4·2 d_x	Deviations of y from the mean 188·6 d_y	d_x^2	d_y^2 by tables of squares	Products of deviations $d_x d_y$
2·6	138	−1·6	−50·6	2·56	2560·0	80·96
3·1	163	−1·1	−25·6	1·21	655·4	28·16
3·5	166	−0·7	−22·6	0·49	510·8	15·82
3·7	153	−0·5	−35·6	0·25	1267·0	17·80
4·1	177	−0·1	−11·6	0·01	134·6	1·16
4·4	201	0·2	12·4	0·04	153·8	2·48
4·6	216	0·4	27·4	0·16	750·8	10·96
4·9	208	0·7	19·4	0·49	376·4	13·58
5·3	226	1·1	37·4	1·21	1399·0	41·14
5·8	238	1·6	49·4	2·56	2440·0	79·04
Total 42·0	1886	0·0	00·0	8·98	10,247·8	291·10

Covariance, S_{xy} = mean of products of deviations

$$= \Sigma d_x d_y / 10$$
$$= 29{\cdot}11.$$

Standard deviation of the 10 values of x,

$$S_x = \sqrt{\{\Sigma d_x^2/10\}}$$
$$= 0{\cdot}9476.$$

Standard deviation of the 10 values of y,

$$S_y = \sqrt{\{\Sigma d_y^2/10\}}$$
$$= 32{\cdot}01.$$

Coefficient of correlation,

$$r_{xy} = S_{xy}/(S_x \, . \, S_y)$$
$$= 29{\cdot}11/(0{\cdot}9476 \times 32{\cdot}01)$$
$$= 0{\cdot}96 \text{ (to two places of decimals).}$$

TV sets, y, of § 102. It emphasizes the need for care with + and − signs when writing down the products of deviations.

114. Terminology. By analogy with mechanics the deviations from the mean, d_x and d_y, are often called first moments about the mean, while d_x^2 and d_y^2 are called *second moments*. A convenient name for $d_x d_y$ is, therefore, *product-moment*, and r_{xy} is known as the *product-moment coefficient of correlation*.

TABLE 9B

Calculation of coefficient of correlation (inverse)

Cinema admissions (millions) x	TV sets (per 1000 population) y	Deviations of x from the mean 9·68 d_x	Deviations of y from the mean 65·08 d_y	By four-figure tables d_x^2	d_y^2	$d_x\,d_y$
11·0	12	1·32	−53·08	1·7420	2818·0	− 70·07
10·0	18	0·32	−47·08	0·1024	2217·0	− 15·07
10·0	24	0·32	−41·08	0·1024	1688·0	− 13·15
9·4	37	−0·28	−28·08	0·0784	788·5	+ 7·86
10·5	52	0·82	−13·08	0·6724	171·1	− 10·73
9·7	64	0·02	− 1·08	0·0004	1·2	− 0·02
9·4	69	−0·28	3·92	0·0784	15·4	− 1·10
9·3	81	−0·38	15·92	0·1444	253·4	− 6·05
9·9	98	0·22	32·92	0·0484	1083·0	+ 7·24
9·3	101	−0·38	35·92	0·1444	1290·0	− 13·65
9·0	106	−0·68	40·92	0·4624	1675·0	− 27·83
8·6	119	−1·08	53·92	1·1660	2907·0	− 58·23
*Total 116·1	781	−0·06	0·04	4·7420	14,907·6	−200·80

Covariance,
$$S_{xy} = \Sigma d_x\, d_y/12 = -16{\cdot}73.$$

Standard deviation,
$$S_x = \sqrt{\{\Sigma d_x^2/12\}} = 0{\cdot}6286.$$

Standard deviation,
$$S_y = \sqrt{\{\Sigma d_y^2/12\}} = 35{\cdot}25.$$

Coefficient of correlation,
$$r_{xy} = S_{xy}/(S_x\, S_y) = -16{\cdot}76/(0{\cdot}6286 \times 35{\cdot}25) = -0{\cdot}76 \text{ (to two places of decimals).}$$

115. Table of minimum numerical values r_{xy}. To establish that correlation exists between two variables, r_{xy} should be calculated from as large a number of pairs of observations as possible. If r_{xy} is calculated from a small number of (x, y) pairs it is possible that, due to pure chance, quite a high numerical value is obtained even though x and y are not correlated. On the other hand if r_{xy} is calculated from a large number of (x, y) pairs chance effect is virtually eliminated and quite a small

* The totals of the d_x and d_y columns would, of course, be zero if the means were exact.

numerical value is an indication that x and y are correlated. Table 9 C will be found useful for making a decision about r_{xy} once it has been calculated. It indicates that if only 6 (x,y) pairs are used the numerical value of r_{xy} (i.e. $|r_{xy}|$) must be at least 0·82 for correlation to be accepted as likely. That is to say, values only slightly smaller than 0·82 may be due to chance and not to correlation. On the other hand when 20 (x, y) pairs are used $|r_{xy}|$ need only be 0·45 for correlation to be accepted as likely.

TABLE 9C

No. of pairs of values of x and y from which r_{xy} is calculated	Minimum value of $\lvert r_{xy}\rvert$ for correlation to be likely
5	0·88
6	0·82
7	0·76
8	0·71
9	0·67
10	0·64
11	0·61
12	0·58
13	0·56
14	0·54
15	0·52
16	0·50
18	0·47
20	0·45
40	0·31
80	0·22
100	0·20

116. Exercises.

1–10. Calculate the product-moment coefficients of correlation of the bivariate distributions given in exercises 1–10, § 110.

11. From the following data calculate the correlation coefficient between the wheat yield and the area planted in ten successive years.

Wheat yield	Area planted
1·3	1·4
1·1	1·4
1·0	1·2
1·1	1·3
1·6	1·7
1·8	1·8
1·6	1·8
1·4	1·7
1·4	1·7
1·9	1·8

(J.M.B.)

12. Compute the coefficient of correlation between the following series of index numbers.

Wholesale prices	Physical volume of production
64	85
63	81
60	81
64	91
63	91
61	83
60	78
61	87
60	88
60	89

What meaning do you attach to your result? (London.)

13. Calculate the coefficient of correlation for the following series:

	Import prices X	Export prices Y
Jan.	223	211
Feb.	226	215
Mar.	229	220
Apr.	234	221
May	241	225
June	245	227
July	252	230
Aug.	254	234
Sept.	254	236
Oct.	256	237

What meaning do you attach to your result? (London.)

14. *Index Numbers of the cost of living* (X) *and wages rates* (Y), *for ten successive years*

X	Y
89	101
89	104
91	106
101	117
114	127
114	136
114	144
115	153
116	161
116	175

Calculate the coefficient of correlation for these series.

What light does your result throw on the relationship between the two series? (London.)

15. *Interim index of production* (X) *and index of volume of exports* (Y)

Year	Quarter	X	Y
1946	1	94	84
	2	98	98
	3	99	104
	4	109	112
1947	1	96	101
	2	109	102
	3	109	114
	4	119	118
1948	1	120	126
	2	122	134

Calculate the coefficient of correlation for these series.
What meaning do you attach to your result? (London.)

16. *Great Britain 1955–67*

Year	Vehicles (millions) x	Casualties (thousands) y
1955	6·4	268
1956	6·9	268
1957	7·5	274
1958	8·0	300
1959	8·7	333
1960	9·4	348
1961	10·0	350
1962	10·6	342
1963	11·4	356
1964	12·4	385
1965	12·9	398
1966	13·3	392
1967	14·1	370

The above data continues the bivariate distribution of § 100. For these 13 (x, y) pairs calculate $(\bar{x}, \bar{y})$ and S_x, S_y, S_{xy}.

Hence obtain r_{xy}. Also by using equation

$$y-\bar{y} = \frac{S_{xy}}{S_x^2}(x-\bar{x})$$

show that the regression of y on x is $y = 17x+167$. Draw a diagram to illustrate this line and the original data.

10

Correlation by ranks

117. Spearman's coefficient of rank correlation ρ. The following table shows the data of § 100 with *ranks* attached to the x and y values:

No. of vehicles (millions) x	Rank R_x of the x values	No. of casualties (thousands) y	Rank R_y of the y values
2·6	10	138	10
3·1	9	163	8
3·5	8	166	7
3·7	7	153	9
4·1	6	177	6
4·4	5	201	5
4·6	4	216	3
4·9	3	208	4
5·3	2	226	2
5·8	1	238	1

The number 2·6, for example, is *tenth* in *order* or *rank* of the x values, 4·1 is *sixth* and 5·8 is *first*. If the product-moment coefficient of correlation is calculated from the ranks R_x and R_y instead of the original values of x and y far less arithmetic is involved and quite a good approximation to r_{xy} is obtained which is denoted by the Greek letter ρ and called the *coefficient of rank correlation.* This idea was originally introduced by C. Spearman in 1906. By applying the method shown in table 9 A to the R_x and R_y values the student will obtain $\rho = 0{\cdot}95$.

118. The formula for the coefficient of rank correlation. The ranks are the numbers 1, 2, 3, 4, ..., n (n in the example under consideration being 10), and the algebraic formulae for the sum of the first n natural numbers and for the sum of their squares can be used to prove that ρ may be calculated as shown in table 10 A by the formula

$$\rho = 1 - \frac{6\Sigma D^2}{n(n^2-1)},$$

where D is the *rank difference.*

TABLE 10A

Calculation of Spearman's coefficient ρ (direct correlation)

Rank of no. of vehicles R_x	Rank of no. of casualties R_y	Rank difference $R_x - R_y$ D	D^2
10	10	0	0
9	8	1	1
8	7	1	1
7	9	−2	4
6	6	0	0
5	5	0	0
4	3	1	1
3	4	−1	1
2	2	0	0
1	1	0	0
	Total	0	8

Since $n = 10$ and $\Sigma D^2 = 8$,

$$p = 1 - \{6 \times 8/10(10^2 - 1)\}$$
$$= 0{\cdot}95.$$

119. Method of ranking equal values of a variate. If the *cinema admissions (millions)* of § 102 are arranged in descending order of magnitude they appear as follows:

11·0, 10·5, 10·0, 10·0, 9·9, 9·7, 9·4, 9·4, 9·3, 9·3, 9·0, 8·6.

The rank of 11·0 is, therefore, 1 and that of 10·5 is 2. The rank of the *two* values 10·0 is not taken as 3 or 4 but as 3·5, the mean of the ranks 3 and 4. Similarly, the ranks of the two values 9·4 is taken as 7·5 and that of the two values 9·3 as 9·5. The twelve ranks are thus:

1, 2, 3·5, 3·5, 5, 6, 7·5, 7·5, 9·5, 9·5, 11, 12

and their sum is 78, the same as the sum of the ranks

1, 2, 3, 4, 5, 6, 7, 8, 9, 10, 11, 12

when the values of the variate are all different.

Table 10 B shows the method of calculating the coefficient of rank correlation for the *cinema admissions*, x, and *TV sets*, y, of § 102. It is a good example of inverse correlation. The value obtained, $\rho = -0{\cdot}84$, is considerably higher (numerically) than the value $r_{xy} = -0{\cdot}75$ calculated in table 9 B. It must be clearly understood that table 9 C is for minimum values of $|r_{xy}|$. Table 10 C is a similar table for minimum values of $|\rho|$ prepared from M. G. Kendall's tables.

TABLE 10B

Calculation of Spearman's coefficient ρ (inverse correlation)

Cinema admissions (millions) x	Rank R_x	TV sets per 1000 population y	TV sets Rank R_y	Rank difference $R_x - R_y$ D	D^2
11·0	1	12	12	−11	121
10·0	3½	18	11	− 7½	56¼
10·0	3½	24	10	− 6½	42¼
9·4	7½	37	9	− 1½	2¼
10·5	2	52	8	− 6	36
9·7	6	64	7	− 1	1
9·4	7½	69	6	1½	2¼
9·3	9½	81	5	4½	20¼
9·9	5	98	4	1	1
9·3	9½	101	3	6½	42¼
9·0	11	106	2	9	81
8·6	12	119	1	11	121
			Total	0	526½

Since $n = 12$ and $\Sigma D^2 = 526\frac{1}{2}$

$$p = 1 - \{6 \times 526\tfrac{1}{2}/12(12^2 - 1)\}$$
$$= -0{\cdot}84.$$

TABLE 10C

No. of pairs from which ρ is calculated n	Minimum value of $\lvert\rho\rvert$ for correlation to be likely
5	1
6	0·89
7	0·79
8	0·74
9	0·68
10	0·65
20	0·46

120. Kendall's coefficient of rank correlation τ. Recently the theory and application of rank correlation methods have been considerably developed by M. G. Kendall in *Rank Correlation Methods*, published in 1948 and revised in 1955 and 1962. Kendall emphasizes that ranks have an importance all their own. They fulfil functions which are more than mere approximations to the (x, y) pairs. Numerical values cannot be attached to *aesthetic appeal, flavour, degree of whiteness, surface polish.* Qualities such as these, however, can be used when obtaining ranks. Kendall also introduces his coefficient of correlation denoted by the Greek letter τ which at first seems quite unrelated to r_{xy} or ρ, as it is calculated in a completely different way. It is interesting that H. E.

Daniels in the *Journal of the Royal Statistical Society*, series B (1950) showed that

$$-1 \leqslant 3\tau - 2\rho \leqslant 1.$$

The full calculation of Kendall's τ for the rankings of vehicles and casualties is set out in table 10 D. For convenience, the order of the years has been reversed so that the rankings R_x appear in ascending order and the rankings R_y are then slightly out of ascending order. Each (R_x, R_y) pair has been lettered in order *A, B, C, D, E, F, G, H, I, J*.

TABLE 10 D

Calculation of Kendall's coefficient τ (direct)

R_x and R_y are reproduced from table 10 A in ascending order for convenience.

	1954	1953	1952	1951	1950	1949	1948	1947	1946	1945	
	A	*B*	*C*	*D*	*E*	*F*	*G*	*H*	*I*	*J*	
R_x	1	2	3	4	5	6	7	8	9	10	Horizontal
R_y	1	2	4	3	5	6	9	7	8	10	totals
Pair		*AB*	*AC*	*AD*	*AE*	*AF*	*AG*	*AH*	*AI*	*AJ*	
Score		+1	+1	+1	+1	+1	+1	+1	+1	+1	+ 9
Pair			*BC*	*BD*	*BE*	*BF*	*BG*	*BH*	*BI*	*BJ*	
Score			+1	+1	+1	+1	+1	+1	+1	+1	+ 8
Pair				*CD*	*CE*	*CF*	*CG*	*CH*	*CI*	*CJ*	
Score				−1	+1	+1	+1	+1	+1	+1	+ 5
Pair					*DE*	*DF*	*DG*	*DH*	*DI*	*DJ*	
Score					+1	+1	+1	+1	+1	+1	+ 6
Pair						*EF*	*EG*	*EH*	*EI*	*EJ*	
Score						+1	+1	+1	+1	+1	+ 5
Pair							*FG*	*FH*	*FI*	*FJ*	
Score							+1	+1	+1	+1	+ 4
Pair								*GH*	*GI*	*GJ*	
Score								−1	−1	+1	− 1
Pair									*HI*	*HJ*	
Score									+1	+1	+ 2
Pair										*IJ*	
Score										+1	+ 1
								Total score, *S*			+39

Number of rankings, $n = 10.$

Maximum possible score, $\frac{1}{2}n(n-1) = 45.$

$$\text{Kendall's } \tau = \frac{S}{\frac{1}{2}n(n-1)}$$
$$= 0{\cdot}87.$$

Consider every pair of letters that can be chosen from these ten letters: *AB, AC, AD, ..., BC, BD, BE, ..., CD, CE, CF,....* Altogether 45 such pairs are possible and in general, if there are *n* letters instead of 10, it can be proved that $\frac{1}{2}n(n-1)$ pairs are possible. The 45 pairs are set out according to an obvious system in table 10 D and to each pair is attached a score, either $+1$ or -1. The way in which this score is obtained is illustrated by the following examples:

To obtain the score for AB

The ranks of R_x are 1, 2. These are in the *right* order and score $+1$.
The ranks of R_y are 1, 2. These are in the *right* order and score $+1$.
The product of the two AB scores is $(+1)(+1) = +1$.

To obtain the score for AG

The ranks of R_x are 1, 7. These are in the *right* order and score $+1$.
The ranks of R_y are 1, 9. These are in the *right* order and score $+1$.
The product of the two AG scores is $(+1)(+1) = +1$.

To obtain the score for CD

The ranks of R_x are 3, 4. These are in the *right* order and score $+1$.
The ranks of R_y are 4, 3. These are in *reverse* order and score -1.
The product of the two CD scores is $(+1)(-1) = -1$.

To obtain the score for GI

The ranks of R_x are 7, 9. These are in the *right* order and score $+1$.
The ranks of R_y are 9, 8. These are in *reverse* order and score -1.
The product of the two GI scores is $(+1)(-1) = -1$.

Finally, table 10 D indicates that the total score, S, is 39 and Kendall defines τ by

$$\tau = \frac{\text{total score}}{\text{maximum possible score}}$$

$$= \frac{S}{\frac{1}{2}n(n-1)}.$$

TABLE 10 E

Method of calculating Kendall's τ when the ranks R_x are in ascending order

R_y	Method of scoring	Score
1	Has 9 larger on its right	+9
2	Has 8 larger on its right	+8
4	1 smaller, 6 larger on its right	+5
3	Has 6 larger on its right	+6
5	Has 5 larger on its right	+5
6	Has 4 larger on its right	+4
9	2 smaller, 1 larger on its right	−1
7	Has 2 larger on its right	+2
8	Has 1 larger on its right	+1
	Total score, S	+39
	Maximum possible score, $\frac{1}{2}n(n-1)$	45

$$\text{Kendall's } \tau = \frac{39}{45}$$

$$= 0{\cdot}87.$$

The complete procedure set out in table 10 D can be shortened considerably. First of all it is evident that the score is +1 or −1 according as the two rankings *agree* or *disagree*. This eliminates the product idea of (+1) (+1), (+1) (−1), (−1) (−1). Secondly, because the R_x rankings are in strict ascending order it is only necessary to consider the R_y rankings and these can be assessed as shown in table 10 E.

TABLE 10F

Calculation of Kendall's τ with tied ranks (inverse)

R_x and R_y are reproduced from table 10 B.

	A	*B*	*C*	*D*	*E*	*F*	*G*	*H*	*I*	*J*	*K*	*L*	Horizontal totals
R_x	1	$3\frac{1}{2}$	$3\frac{1}{2}$	$7\frac{1}{2}$	2	6	$7\frac{1}{2}$	$9\frac{1}{2}$	5	$9\frac{1}{2}$	11	12	
R_y	12	11	10	9	8	7	6	5	4	3	2	1	
Pair		*AB*	*AC*	*AD*	*AE*	*AF*	*AG*	*AH*	*AI*	*AJ*	*AK*	*AL*	
Score		−1	−1	−1	−1	−1	−1	−1	−1	−1	−1	−1	−11
Pair			*BC*	*BD*	*BE*	*BF*	*BG*	*BH*	*BI*	*BJ*	*BK*	*BL*	
Score			0	−1	+1	−1	−1	−1	−1	−1	−1	−1	− 7
Pair				*CD*	*CE*	*CF*	*CG*	*CH*	*CI*	*CJ*	*CK*	*CL*	
Score				−1	+1	−1	−1	−1	−1	−1	−1	−1	− 7
Pair					*DE*	*DF*	*DG*	*DH*	*DI*	*DJ*	*DK*	*DL*	
Score					+1	+1	0	−1	+1	−1	−1	−1	− 1
Pair						*EF*	*EG*	*EH*	*EI*	*EJ*	*EK*	*EL*	
Score						−1	−1	−1	−1	−1	−1	−1	− 7
Pair							*FG*	*FH*	*FI*	*FJ*	*FK*	*FL*	
Score							−1	−1	+1	−1	−1	−1	− 4
Pair								*GH*	*GI*	*GJ*	*GK*	*GL*	
Score								−1	+1	−1	−1	−1	− 3
Pair									*HI*	*HJ*	*HK*	*HL*	
Score									+1	0	−1	−1	− 1
Pair										*IJ*	*IK*	*IL*	
Score										−1	−1	−1	− 3
Pair											*JK*	*JL*	
Score											−1	−1	− 2
Pair												*KL*	
Score												−1	− 1
										Total score, *S*			−47

Number of rankings, $n = 12.$

Maximum possible score, $-\frac{1}{2}n(n-1) = 66.$

$$\text{Kendall's } \tau = \frac{S}{\frac{1}{2}n(n-1)}$$

$$= -0{\cdot}71.$$

Table 10 E is so much shorter than table 10 D that it is evidently a good idea to put the ranks R_x in ascending order even though they are not in order when first obtained. Once the method of scoring is understood it need not be recorded as part of the calculation of τ.

Table 10 F shows how tied ranks produce *zeros* in the scores. One example will be sufficient to illustrate the principle:

To obtain the score for BC

The ranks of R_x are $3\frac{1}{2}$, $3\frac{1}{2}$. These are *equal* and score 0.
The ranks of R_y are 11, 10. These are in *reverse* order and score -1.
The product of the two *BC* scores is $(0)(-1) = 0$.

Table 10 G has been prepared from M. G. Kendall's tables. It will be found useful for making a decision about τ once it has been calculated.

TABLE 10G

No. of pairs from which τ is calculated, n	Minimum value of $\lvert\tau\rvert$ for correlation to be likely
5	1
6	0·87
7	0·71
8	0·64
9	0·56
10	0·51
20	0·33

121. Exercises.

1–15. Calculate Spearman's ρ and Kendall's τ for the bivariate distributions of exercises 1–15, § 116. Note that they are approximately equal to their respective product-moment coefficients of correlation.

16. Ten shades of the colour green when arranged in their true order from light to dark are numbered 1–10 respectively. An observer, when asked to arrange the shades from light to dark, produces the following rank

3, 1, 5, 2, 6, 4, 10, 9, 7, 8.

What is the value of Spearman's coefficient of rank correlation in this case? (London.)

17. Explain briefly what is meant by correlation.

A class of 15 students was examined in Mathematics and Physics. The following table gives the orders of merit of the students (arranged in alphabetical order) in both subjects. Calculate the coefficient of ranked correlation:

Mathematics	11	5	9	14	13	2	6	10	1	4	7	3	12	8	15
Physics	11	8	7	9	12	1	4	10	3	6	14	2	13	5	15

(London.)

18. There are ten finalists in a competition for which there are two judges *X* and *Y*. The following table gives the order in which *X* and *Y* place the

competitors. Calculate the rank correlation coefficient and state any conclusions that you deduce from it:

Competitors	*A*	*B*	*C*	*D*	*E*	*F*	*G*	*H*	*K*	*L*
Ranking of *X*	6	10	2	7	1	4	5	9	8	3
Ranking of *Y*	2	7	8	10	6	1	9	5	3	4

(London.)

19. In a drama competition ten plays were ranked by two adjudicators as follows:

Play	*A*	*B*	*C*	*D*	*E*	*F*	*G*	*H*	*J*	*K*
Rank given by *X*	5	2	6	8	1	7	4	9	3	10
Rank given by *Y*	1	7	6	10	4	5	3	8	2	9

Calculate the coefficient of ranked correlation. Is there any reason for saying that there is a significant agreement between the two adjudicators?

(London.)

20. Ten students had a test in Physics and Mathematics. Their scores per cent were:

Physics	100	80	60	57	90	73	68	83	67	56
Mathematics	80	90	66	72	95	64	44	83	74	55

Calculate Spearman's coefficient of rank correlation for these scores. What deductions, if any, can be made about the relation between the students' performances in Physics and Mathematics? (W.A.E.C.)

11

The analysis of a time-series

122. The graph of a time-series, or historigram. A time-series is a series of values assumed by a variable at different points of time. The *cinema admissions*, x, of § 102 is a good example of such a series. Noticeable features of the cinema admission figures are:

(i) For the first quarter of each year they are high compared with the other three quarters. This is an example of *seasonal variation* in the series.

(ii) During the three years they decrease and we say that the *general trend* of the series is downwards.

The graph of the time-series shown in fig. 32 indicates the seasonal variation of the *peaks* P_2, P_3, and the *troughs* T_1, T_2. There is no justification for saying that P_1 is a peak and T_3 a trough unless the preceding and following figures are known. It gives the *history* of the cinema admissions during the period 1950–2 and is called the *historigram.* The *four-quarterly moving averages* graph included in the same diagram indicates the general trend.

123. The four-quarterly moving averages. If

$m_1 =$ the mean of the cinema admissions for quarters 1, 2, 3 and 4,
$m_2 =$ the mean of the cinema admissions for quarters 2, 3, 4 and 5,
$m_3 =$ the mean of the cinema admissions for quarters 3, 4, 5 and 6,
etc.,

then m_1, m_2, m_3, etc., are called the four-quarterly moving averages of the cinema admissions. The smoother graph obtained by plotting the moving averages gives a better idea of the general trend than the graph of the unmodified series.

124. A convenient method of calculating the moving averages. In the example under consideration

$$\begin{aligned} m_1 &= \tfrac{1}{4}(11{\cdot}0+10{\cdot}0+10{\cdot}0+9{\cdot}4) \\ &= 10{\cdot}1. \end{aligned}$$

On the graph it is plotted in the middle of the 1–4 quarter range.

Instead of calculating m_2 by $m_2 = \frac{1}{4}(10{\cdot}0+10{\cdot}0+9{\cdot}4+10{\cdot}5)$ it is more convenient to use

$$\begin{aligned} m_2 &= m_1+\tfrac{1}{4}(10{\cdot}5-11{\cdot}0) \\ &= 10{\cdot}1-0{\cdot}125 \\ &= 9{\cdot}975. \end{aligned}$$

Similarly $m_3 = m_2+\frac{1}{4}(9{\cdot}7-10{\cdot}0)$, and so on for m_4, m_5, etc. Table 11 A shows a convenient method of setting out the whole calculation. The *table of differences* is obtained by subtracting each x value from the x value directly below it. The differences are then divided by 4 and

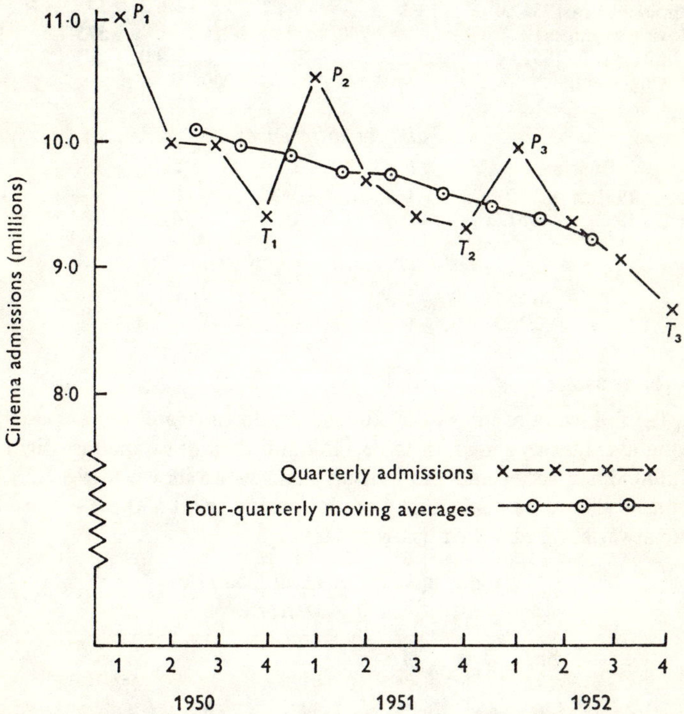

Fig. 32. Historigram of the unmodified series and the four-quarterly moving averages. (The zigzag in the ordinate scale indicates that it does not begin at zero.)

entered in the third section of the table. The values in this third section are then *added cumulatively* to the mean of the first four quarters, 10·100, to obtain the series of four-quarterly moving averages:

10·100, 9·975, 9·900, 9·750, 9·725, 9·575, 9·475, 9·375, 9·200.

These four-quarterly moving averages are finally written in the first section of the table, as shown, in the middle of their respective four-quarterly ranges. Note particularly that the moving averages are entered in the first section of the table after the second and third sections have been completed.

TABLE 11A

Calculation of four-quarterly moving averages

Cinema admissions, x

Quarter...	1		2		3		4	
Admissions in 1950, x	11·0		10·0		10·0		9·4	
Moving averages				10·100		9·975		9·900
Admissions in 1951, x	10·5		9·7		9·4		9·3	
Moving averages		9·750		9·725		9·575		9·475
Admissions in 1952, x	9·9		9·3		9·0		8·6	
Moving averages		9·375		9·200				

Table of differences

Quarter... ...	1	2	3	4
1951, $x-$1950, x	−0·5	−0·3	−0·6	−0·1
1952, $x-$1951, x	−0·6	−0·4	−0·4	−0·7

Differences divided by four

−0·125	−0·075	−0·150	−0·025
−0·150	−0·100	−0·100	−0·175

125. Exercises.

In the following exercises the student should calculate the four-quarterly moving averages as shown in table 11 A and then draw the historigram of the unmodified series and the moving averages as shown in fig. 32. Brief comments should be made finally on the general trend and the extent of the seasonal variation above or below the trend.

1. *Index numbers of retail food prices*

Year	Quarter 1	2	3	4
1968	72	71	76	76
1969	75	74	77	80
1970	80	81	83	85

(London.)

2. *Quarterly index numbers of coal production*

Year	Quarter 1	2	3	4
1968	93·3	81·7	81·5	89·1
1969	93·8	92·3	86·5	93·7
1970	97·6	82·2	79·0	89·3

(London.)

3. *Percentages of workpeople unemployed*

Year	Quarter 1	2	3	4
1965	11·2	10·9	12·0	11·0
1966	10·4	14·3	14·0	13·5
1967	10·9	8·8	9·3	9·8
1968	10·4	9·8	11·5	11·7

(London.)

Note the *abnormal* rise in 1966 is not a seasonal variation.

4. *Production of motor vehicles (tens of thousands)*

Year	Quarter 1	2	3	4
1967	26	36	24	11
1968	29	36	36	22
1969	40	52	43	17

(London.)

5. *Expenditure on fuel and light*

Year	Quarter 1	2	3	4
1966	84	64	61	82
1967	92	70	63	85
1968	100	81	72	96

(London.)

6. *Number of licences current for road vehicles in Great Britain at the end of the month stated (tens of thousands)*

	1947	1948	1949	1950	1951	1952
Feb.	295	261	350	383	393	416
May	329	315	386	419	436	462
Aug.	344	365	402	431	452	477
Nov.	325	368	405	432	449	475

(London.)

126. The twelve-monthly moving averages. Table 11 B shows the monthly production figures of the manufacturing industries of the United Kingdom. Seasonal variations in this series recur at twelve-monthly intervals. They are eliminated by the twelve-monthly moving averages shown in the same table. The calculation shown in table 11 B is an obvious extension of that of table 11 A to groups of twelve instead of four. The histogram (fig. 33), shows that the seasonal variation takes the form of a definite drop in production each year during the summer holiday period. The moving average graph shows a steady increase in the general trend of about 10 points per annum (i.e. 10 % of the production of 1946).

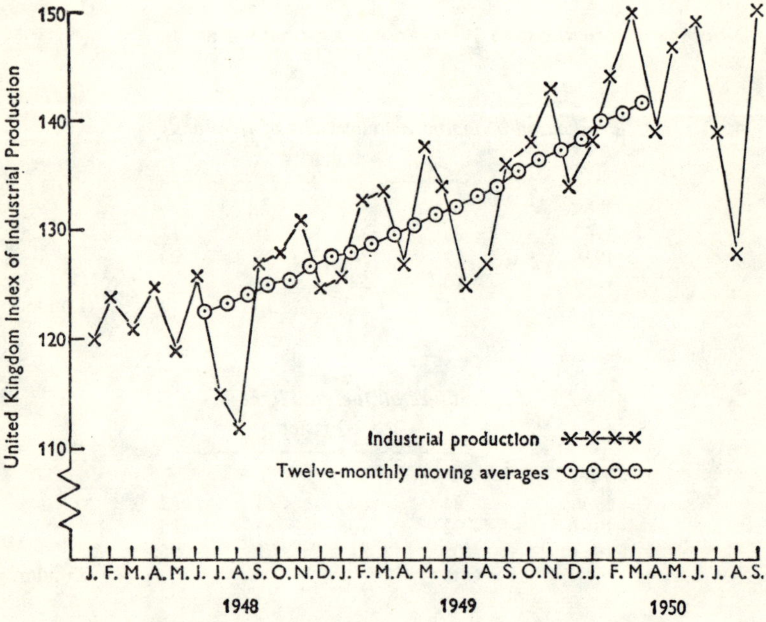

Fig. 33. Historigram of the monthly index of industrial production and the 12-monthly moving averages.

127. The seven-day moving averages. In the daily attendances at an exhibition given in table 11 C a rise and fall due to the day of the week is noticeable. On Sunday the attendance is low, on Monday it is high and on Tuesday it is low again. During the rest of the week the attendances steadily increase until the maximum is reached on Saturday. These same fluctuations recur in the second week and they are eliminated by the seven-day moving average which is calculated in table 11 C and illustrated graphically in fig. 34. The general trend shows a very slight increase in attendances during the two weeks.

128. The ten-yearly moving averages. Table 11 D shows the total number of births in the United Kingdom each year for a period of 22 years and

TABLE 11B

Calculation of twelve-monthly moving averages

Index of industrial production, 1946 = 100 (all manufacturing industries)

	Jan.	Feb.	Mar.	Apr.	May	June	July	Aug.	Sept.	Oct.	Nov.	Dec.
1948 indices	120	124	121	125	119	126	115	112	127	128	131	125
Moving averages						122·75	123·25	124·00	125·08	125·25	126·83	127·50
1949 indices	126	133	134	127	138	134	120	122	136	138	143	134
Moving averages	127·92	128·75	129·50	130·33	131·33	132·08	133·08	134·00	135·33	136·33	137·08	138·33
1950 indices	138	144	150	139	147	149	139	128	150			
Moving averages	139·92	140·42	141·58									

Table of differences

	Jan.	Feb.	Mar.	Apr.	May	June	July	Aug.	Sept.	Oct.	Nov.	Dec.
1949 – 1948	6	9	13	2	19	8	5	10	9	10	12	9
1950 – 1949	12	11	16	12	9	15	19	6	14			

Differences divided by twelve (to two decimals)

Jan.	Feb.	Mar.	Apr.	May	June	July	Aug.	Sept.	Oct.	Nov.	Dec.
0·50	0·75	1·08	0·17	1·58	0·67	0·42	0·83	0·75	0·83	1·00	0·75
1·00	0·92	1·33	1·00	0·75	1·25	1·58	0·50	1·17			

TABLE 11C

Calculation of seven-day moving averages

Daily attendances, in thousands, at an exhibition

	Sun.	Mon.	Tues.	Wed.	Thurs.	Fri.	Sat.
1st week	35	70	36	59	62	68	71
Moving averages				57·29	57·86	58·14	58·43
2nd week	39	72	38	56	63	71	75
Moving averages	58·00	58·14	58·57	59·14			

Table of differences

2nd week – 1st week	4	2	2	–3	1	3	4

Differences divided by seven (to two decimals)

0·57	0·29	0·29	–0·43	0·14	0·43	0·57

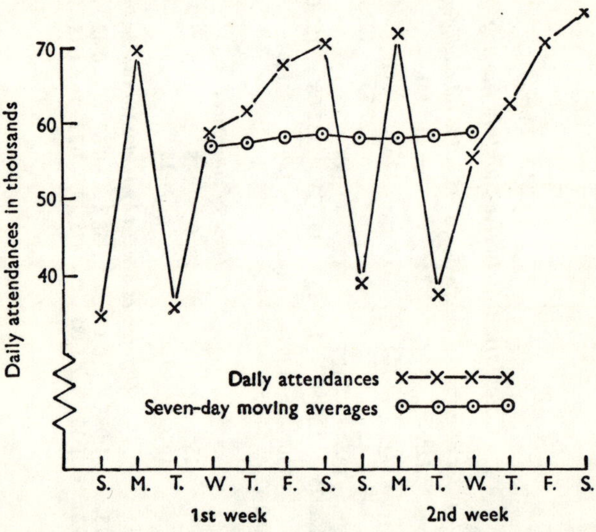

Fig. 34. Daily attendances and the seven-day moving averages.

fig. 35 is the historigram. Notice that there is a rise and fall in the period 1933–41, a second rise and fall in the period 1941–5 and a third rise and fall in the period 1946–52. These are examples of *cyclical variations*. An upward trend over a period of twenty or thirty years is usually attained, not by a steady increase each year, but by cycles which often vary in period from five to ten years. The ten-yearly moving averages, which are calculated in table 11 D and illustrated graphically

TABLE 11D

Calculation of ten-yearly moving averages

Total number of births, in thousands, in the United Kingdom between the years 1933 and 1954

	1933		1934		1935		1936		1937		1938		1939		1940		1941		1942	
Births	692		712		711		720		724		736		727		702		696		772	
Moving averages										719·2		731·1		747·7		756·2		779·7		809·8
	1943		1944		1945		1946		1947		1948		1949		1950		1951		1952	
Births	811		878		796		955		1025		905		855		818		797		793	
Moving averages		826·7		839·5		851·1		861·2		863·3		862·2		854·3						
	1953		1954																	
Births	804		795																	
Moving averages																				

Table of differences

119	166	85	235	301	169	128	116	101	21
−7	−83								

Differences divided by ten

11·9	16·6	8·5	23·5	30·1	16·9	12·8	11·6	10·1	2·1
−0·7	−8·3								

in fig. 35, remove the cyclical variations and reveal an upward trend in the birth-rate until 1947 and the beginning of a decline after 1947. The ten-yearly moving averages are effective in this case because the complete fluctuation in the birth-rate occurs over a period which is roughly ten years. One would expect seven-yearly moving averages to be effective for cyclical variations of period seven years.

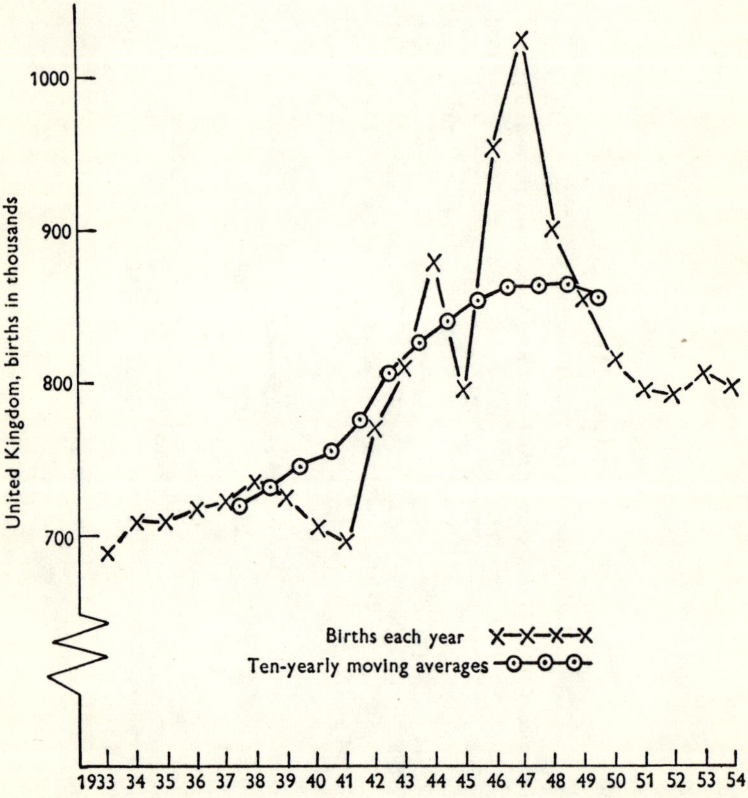

Fig. 35. The rising birth-rate of the United Kingdom.

129. Summary of the analysis of a time-series. In the analysis of a time-series it is necessary to distinguish between the following types of variation:

(i) *The general trend* over several decades. It is also called the *secular variation.*

(ii) *Cyclical variations* which are fluctuations in the general trend.

Their periods are often between five and ten years. It is generally possible to eliminate them by a suitably chosen moving average.

(iii) *Seasonal variations* which recur annually due to the seasons of the year. They are eliminated by the four-quarterly or twelve-monthly moving average.

(iv) *Special variations* due to abnormal circumstances. The high unemployment figures in the second and third quarters of 1966 (see § 125, exercise 3) and the high-birth rate in 1947 after the *Second World War* (see table 11 D) are examples of special variation.

An example has also been given of variations due to the day of the week. These were removed by taking seven-day moving averages. It is equally possible that, in a time-series consisting of weekly figures, variations may occur due to the week of the month. Such variations could be conveniently removed by four-weekly or five-weekly moving averages.

130. Exercises.

1. Explain what is meant by 'moving average'.

The following table gives the numbers (in 1000's) of passenger cars and chassis produced for export in the months from April 1948 to June 1950. Construct a table showing the twelve-monthly moving averages and draw a graph to show the trend:

	J	F	M	A	M	J	J	A	S	O	N	D
1948	.	.	.	20	20	25	18	16	25	18	17	19
1949	18	19	26	18	21	26	16	20	26	24	28	28
1950	32	33	41	29	34	40	.	.	.	.	.	.

(London.)

2. The table below shows the number of houses completed each month from January 1952 to August 1953, the numbers being in thousands:

	J	F	M	A	M	J	J	A	S	O	N	D
1952	13·5	14·6	19·0	15·4	17·3	17·0	18·2	15·6	19·7	20·5	19·5	19·7
1953	17·8	17·5	25·0	20·2	22·8	22·5	24·9	24·7	.	.	.	.

Plot these figures on a graph. Calculate the twelve-monthly moving averages and plot its graph with your previous one. (London.)

3. A Ballet Company gave a six weeks' season at a large hall capable of seating 4000 people and the attendances in hundreds, at the evening performances, are recorded in the following table:

	Attendances, in hundreds, to nearest hundred					
	Mon.	Tues.	Wed.	Thurs.	Fri.	Sat.
First week	12	13	20	19	17	24
Second week	22	25	31	31	26	34
Third week	30	31	40	38	36	40
Fourth week	40	40	40	40	40	40
Fifth	38	39	40	40	40	40
Sixth week	32	33	34	32	30	36

Plot a graph of the above time-series and include on the same diagram the graph of the *six-day* moving averages.

Comment on the weekly *cycle* of attendances and state, with reasons, if you think an extension of the season to eight weeks would have been justified. (J.M.B.)

4. Draw roughly (not on graph paper) a diagram to represent a series of twelve annual figures for the price of a commodity, satisfying the following conditions: the general tendency during the period was for the price to rise, but for four separate years the price was lower than for the preceding year. Point out the features of your graph that satisfy the conditions. (J.M.B.)

5. Give an account of the analysis of a time-series indicating why the ordinary statistical averages are not suitable in this case. Explain the use of the method of moving averages to isolate the general trend of such a series. (London.)

6. What do you understand by cyclical variations in trade and business?

As a manager of a sales organization, describe briefly how you would estimate next year's sales. You may assume, for illustrative purposes, that your sales organization deals with any well-known commodity or type of commodity. (London.)

7. Plot a graph showing the changes with time in the following index numbers of real wages and comment on the nature of the fluctuations:

Year	Index no.	Year	Index no.
1880	75	1890	95
1881	78	1891	93
1882	78	1892	90
1883	81	1893	90
1884	78	1894	92
1885	87	1895	95
1886	79	1896	99
1887	84	1897	98
1888	88	1898	98
1889	91	1899	102

Calculate the ten-yearly moving averages and plot them on the same diagram. Comment on the general trend or real wages during this period. (J.M.B.)

8. From the following data calculate the five-yearly moving averages of the prices of beef and mutton for the years 1925–38 and plot their two graphs in one diagram.

Comment on any correlation in the trend of prices of the two commodities.

Year	Price: Beef	Price: Mutton
1925	80	106
1926	74	89
1927	70	86
1928	74	92
1929	71	89
1930	73	92
1931	67	79
1932	65	63
1933	61	69
1934	58	74
1935	54	75
1936	54	73
1937	61	78
1938	62	62

(J.M.B.)

12

Weighted averages

131. Percentage relatives. If the prices of beef and mutton given in § 130, exercise 8, are expressed as percentages of the price in 1925 they are called *percentage relatives with 1925 as the base year.* Table 12 A shows the complete time-series transformed into percentage relatives (to the nearest whole number). The rapid drop in the price of mutton compared with that of beef is much more apparent from the percentage relatives than from the actual prices. The percentage relatives of prices are often called *price relatives.*

TABLE 12A

	Percentage relatives with 1925 as base	
Year	Beef	Mutton
1925	100	100
1926	93	84
1927	88	81
1928	93	87
1929	89	84
1930	91	87
1931	84	75
1932	81	59
1933	76	65
1934	73	70
1935	68	71
1936	68	69
1937	76	74
1938	78	58

132. Weighted average. If we wish to study the variation in the price of meat in general we might take the average of 93 and 84 (i.e. $88\frac{1}{2}$) as an indication of the price in 1926. Research throughout the United Kingdom shows, however, that for every 24 kg of beef consumed only 14 kg of mutton are consumed. This means that an alteration in the price of mutton will not have such a great effect on the amount of money spent as an alteration in the price of beef. In fact it seems obvious that to get a true indication of the change in the price of meat we should multiply the price of beef by the weight of beef consumed and the price of mutton by the weight of mutton consumed and divide the sum by the total weight. Thus for 1926 we should get

$$\frac{(93\times 24+84\times 14)}{(24+14)}=\frac{3408}{38}$$

$$=89{\cdot}7.$$

The result of 89·7 thus obtained is called the *weighted average* of the beef and mutton price relatives. It is a *composite* price relative for beef and mutton.

A complete indication of the change in price of meat, however, should also take into account the change in price of pork, sausages, pies, etc., and table 12 B reproduces the weights assigned by the Cost of Living Advisory Committee based on the consumption in 1953–4. Thus if b, m, p and s are the respective price relatives of beef, mutton, pork and sausages, etc., the weights given in table 12 B mean that the weighted average of b, m, p and s is

$$\frac{(24b+14m+6p+25s)}{(24+14+6+25)}.$$

TABLE 12B

	Weight
Beef	24
Mutton and lamb	14
Pork	6
Sausages, pies, canned meat, offal, poultry, etc.	25

133. The arithmetical calculation of a weighted average. The calculation of a weighted average can be written out in exactly the same way as the calculation of the mean of a frequency distribution shown in § 28, the *percentage relatives* and the *weights* taking the place respectively of the *mid-interval weights in kg* and the *frequencies*. If the price relatives of beef, mutton, pork and sausages, etc., in 1926 are 93, 84, 88 and 82 respectively it would be convenient to work with 84 as origin and to tabulate the calculation as shown in table 12 C.

134. Index numbers. The weighted average (or composite price relative) 86·8 obtained in table 12 C is an example of an *index number*. It indicates that the price of meat in 1926 is 86·8 % of the price in the base year 1925. Alternatively, we might say that the price in 1926 was 13·2 % below the price in 1925. The number 86·8 might be called the *price index* of meat. Further examples of index numbers will be given in this chapter. Indeed, some were given in chapter 11. The retail food price index and the coal production index were used in § 125, exercises 1 and 2, and the index of industrial production (with 1946 as base year) was used in § 126, table 11 B.

TABLE 12C

Calculation of a weighted average

	Price relative	Weight w	Price relative with 84 as origin x	wx
Beef	93	24	9	216
Mutton and lamb	84	14	0	0
Pork	88	6	4	24
Sausages, pies, canned meat, offal, poultry, etc.	82	25	−2	−50
	Total	69	Total	190

$$\text{Weighted average} = 84 + \Sigma wx/\Sigma w = 84 + 190/69 = 86{\cdot}8.$$

135. The choice of base. It is always difficult to decide which year will prove the most satisfactory base. For the *cost-of-living index* (an index of the greatest interest to all of us) conditions prior to the outbreak of the First World War formed the standard of reference for the U.K. for many years, July 1914 being taken as the base. By the end of the Second World War this cost-of-living index had risen to 203, that is to say, prices in 1947 were more than twice what they were in 1914. Moreover, it was realized that the method of calculating the cost-of-living index was completely out of date (see § 136). So in June 1947 it was replaced by the *interim index of retail prices*, June 1947 being taken as base. The word 'interim' was used because the new index was a temporary measure to be used until a permanent measure, based on research carried out by the Cost of Living Advisory Committee could be established. This permanent *index of retail prices* was finally established in 1956 and January 1956 was taken as base. By October 1956 this new index had risen to 102·7. If June 1947 had continued as base it would have been 158, and if July 1914 had continued as base it would have been well over 300. Thus, the cost of living in October 1956 was 2·7% above January 1956, 58% above June 1947 and over 200% above July 1914. This example serves to show the way in which the base for an index number was chosen and altered later when it was found necessary to do so.

136. The index of retail prices. The calculation of the U.K. index of retail prices is shown in table 12 D. The price relative 101·8 for food indicates that the price of food on 16 October 1956 was 1·8% above the price on 17 January 1956. The number 101·8 is calculated by extending a table such as table 12 C to include not only meats but all items of food such as bread, flour, bacon, fish, butter, cheese, milk and eggs. The price relative 104·5 for housing is calculated by considering the rises in rent and rates, charges for repairs, etc.; the 103·5 for miscellaneous goods by considering the price of books, newspapers, stationery, medicines, toilet requisites, soap, fancy goods and toys; the 103·1 for transport and vehicles by considering railway fares, bus fares, cost of cycles, etc.; and the 105·6 for services by considering postages, entertainments, laundry, hairdressing, etc.

TABLE 12D

Calculation of the U.K. index of retail prices for 16 October 1956 (17 January 1956 = 100)

Group	Price relative 16 Oct. 1956	Weight w	Price relative with 100 as origin x	wx
Food	101·8	350	1·8	630
Alcholic drink	102·6	71	2·6	184·6
Tobacco	105·3	80	5·3	424
Housing	104·5	87	4·5	391·5
Fuel and light	102·4	55	2·4	132
Durable household goods	101·3	66	1·3	85·8
Clothing and footwear	101·0	106	1·0	106
Transport and vehicles	103·1	68	3·1	210·8
Miscellaneous goods	103·5	59	3·5	206·5
Services	105·6	58	5·6	324·8
	Total	1000	Total	2696·0

$$\begin{aligned}\text{Index of retail prices} &= 100+\Sigma wx/\Sigma w\\ &= +2696/1000\\ &= 102{\cdot}7 \text{ (1 decimal).}\end{aligned}$$

The weights of the second column need some explanation because they are not really weights but estimates of the proportional importance of each group. They might be interpreted by saying that out of every thousand pence 350 are spent on food, 71 on alcoholic drink, etc. Alternatively, it can be said that out of every £10 the amounts spent respectively on food, drink, tobacco, housing are £3·50, £0·71, £0·80, £0·87. The weights result from research into the expenditure of a *random sample* (see § 146) of households of the United Kingdom. They are called 'weights' because the calculation of the index of retail prices as a weighted average of the price relatives is an extension of the simple idea developed in § 132 in which the relative importance of the changes in price of beef, mutton and pork is dependent on the weights of each type of meat consumed.

It has already been mentioned that the cost-of-living index which had been in operation since July 1914 was completely out-of-date in 1947. The reason for this was that people were spending their money differently in 1947. New kinds of food, new entertainments, better forms of transport and completely different fashions and materials for clothing had transformed their way of life. Since 1947 the amounts of money spent on the different groups in the left-hand column of table 12 D have steadily altered and table 12 E shows how the weights were adjusted by the Cost of Living Advisory Committee in the interim period between 1947 and 1962. Since 1962 the weights have been adjusted each year. Those used in 1968 and 1969 are given in Exs. 31 and 32 chapter 15. Also full details of the calculation of the price index for food are given in Ex. 33.

TABLE 12E

The adjustment, between 1947 and 1962, of the weights used in the calculation of the index of retail prices

Group	Weights used from 17 June 1947 to 15 Jan. 1952	Weights used from 15 Jan. 1952 to 17 Jan. 1956 based on estimates of consumption in 1950	Weights used from 17 Jan. 1956 to 16 Jan. 1962 based on ascertained consumption in 1953–4
Food	348	399	350
Alcoholic drink	217 (Alcoholic drink and Tobacco)	78	71
Tobacco		90	80
Housing	88	72	87
Fuel and light	65	66	55
Durable household goods	71	62	66
Clothing and footwear	97	98	106
Transport and vehicles	79 (Transport and vehicles and Services)	91 (Transport and vehicles and Services)	68
Services			58
Miscellaneous	35	44	59

137. The crude and standardized death-rates. In the U.K., the death-rate of a town for any particular year is the *number of deaths per thousand population.* It is usually called the *crude death-rate* because it does not take into account the age distribution of the population. In a town where a large number of elderly retired people reside, the crude death-rate is bound to be high compared with that of a newly developed industrial town whose population usually consists of young married couples and their children. In order to make a fair comparison of the death-rates of all types of town the *standardized* (or *corrected*) *death-rate* is used. This is a weighted average in which the weights bear a definite relation to the age distribution of the population of the whole country. Table 12 F shows its method of calculation. In the first two columns the population of the town is classified according to age, and in the third column the number of deaths in each age group is recorded. The crude death-rate of the town is simply its total number of deaths divided by its total population in thousands. In the fourth column we have the STANDARD age distribution, in this case the percentage of the total population of the whole country in each age group. From the second and third columns the death-rate of each age group is calculated and recorded in the fifth column. These are often called the AGE-SPECIFIC DEATH-RATES. The standardized death-rate is the weighted average of the age-specific death-rates using the standard age distribution as weights.

TABLE 12F

Calculation of the crude and standardized death-rate

Age group	Population p	Deaths in age group d	Standard % of total population of the U.K. in each age group S	Age-specific death-rate. Death-rate of each age group 1000 d/p R	SR
0–5	1000	13	12·1	13·00	157·30
5–15	1250	1	17·7	0·80	14·16
15–25	600	2	15·8	3·33	52·61
25–35	900	3	15·1	3·33	50·28
35–45	800	3	14·0	3·75	52·50
45–55	750	8	11·6	10·67	123·77
55–65	500	12	7·7	24·00	184·80
Over 65	400	20	6·0	50·00	300·00
Total	6200	62	100·0	—	935·42

$$\text{Crude death-rate} = \frac{\text{Total number of deaths}}{\text{Total population in thousands}}$$

$$= 10 \text{ (deaths per thousand population).}$$

$$\text{Standardized death-rate} = \Sigma SR/\Sigma S$$

$$= 9{\cdot}35 \text{ (deaths per thousand population).}$$

In exercise 8, § 138, the death-rate in 1971 is compared with the death-rate in 1951 by taking the age distribution in 1951 as the standard.

138. Exercises.

1. A composite index number is to be constructed from the following index numbers weighted as shown. Using a working mean of 140, or otherwise, calculate the composite index number as an arithmetic mean:

Index numbers	172	166	150	135	130
Weights	1	2	4	6	3

(London.)

2. What is meant by 'weighting' in the construction of index numbers? Using a working mean of 115, or otherwise, calculate as an arithmetic mean correct to the nearest whole number the 'interim index of retail prices' for 18 April 1950, using the following table of price relatives and weights:

	Price relative	Weight
Food	122·0	348
Rent and rates	101·3	88
Clothing	118·4	97
Fuel and light	115·2	65
Household durable goods	110·6	71
Miscellaneous goods	113·3	35
Services	106·6	79
Drink and tobacco	107·5	217

(London.)

3. Explain what is meant by an index number.

Calculate as an arithmetic mean correct to the nearest integer a cost-of-living index from the following table of price relatives and weights:

	Price relative	Weight
Food	122	35
Rent	101	9
Clothing	118	10
Fuel	115	7
Miscellaneous	108	39

(London.)

4. The following table gives the index numbers of three commodities in 1950, taking 1948 as base:

	1948	1950
Food	100	111
Fuel and light	100	105
Clothing	100	106

(*a*) Calculate the weighted average of these index numbers for 1950, when food, fuel and light, and clothing are given weights of 5, 1 and 2 respectively.

(*b*) Calculate, each to the nearest whole number, the 1948 index numbers referred to 1950 as base. (London.)

5. Explain what is meant by a weighted mean. Why is it used instead of a simple mean?

A candidate obtains the following percentage marks in an examination:

Mathematics 72, French 52, History 67, English 59, Physics 82, Geography 76, Chemistry 77.

It is agreed to give treble weight to Mathematics, English and French, and double weight to Physics and Chemistry. What is the candidate's weighted mean percentage? (London.)

6. *Index of retail prices, 21 June 1966*

	Price index 16 Jan. 1952 = 100)	Weight
Food	118·4	298
Alcoholic drink	119·0	67
Tobacco	120·8	77
Housing	129·5	113
Fuel and light	119·5	64
Durable household goods	106·5	57
Clothing and footwear	109·6	91
Transport and vehicles	109·9	116
Miscellaneous	112·3	61
Services	119·5	56

Using the given weights, calculate the weighted arithmetic mean of the above price indices.

7. Distinguish between the crude and the standardized death-rates. Calculate their values from the following table, in which the last row gives the assumed standard age distribution per thousand:

Age group	0–9	10–39	40–59	60–
Population	20,000	60,000	15,000	5000
Deaths in 1953	250	300	350	400
Standard distribution	250	500	200	50

(London.)

8. *Population and deaths of infants under 5 years old*
(in thousands)

	Under 2 years old		2 years and under 5 years		Total	
Year	Population	Deaths	Population	Deaths	Population	Deaths
1951	1195	48·9	1796	8·4	2991	57·3
1971	1367	22·0	2355	2·4	3722	24·4

Calculate crude death-rates for the infant population in both years. Calculate age-specific death-rates, and hence the standard death-rate for 1971, taking the infant population of 1951 as standard.

Describe in words the most prominent features of the data revealed by your analysis. (J.M.B.)

9. The following table shows the base (1963) and current (1970) prices of five staple foods. Using the given weights, calculate the 1970 index for staple foods with 1963 as base.

Item	Base (1963) prices	Current (1970) prices	Weight
Cassava	10	15	30
Rice	5	10	25
Yam	20	24	20
Plantain	4	5	10
Cocoyam	5	7	15
		Total	100

(W.A.E.C.)

13

The presentation and collection of data

139. Diagrammatic representation of statistical data. The student is already familiar with the construction and use of histograms, cumulative frequency curves, scatter diagrams and historigrams. In this chapter various other methods of presenting statistical data in pictorial or diagrammatic forms will be described and discussed.

140. The picturegram or pictogram. *The diagram below is an attempt to show in pictorial form the relative amounts of wheat obtained from different sources and consumed in the United Kingdom in a certain year. The actual amounts are in the ratio 7 : 2 : 1.*

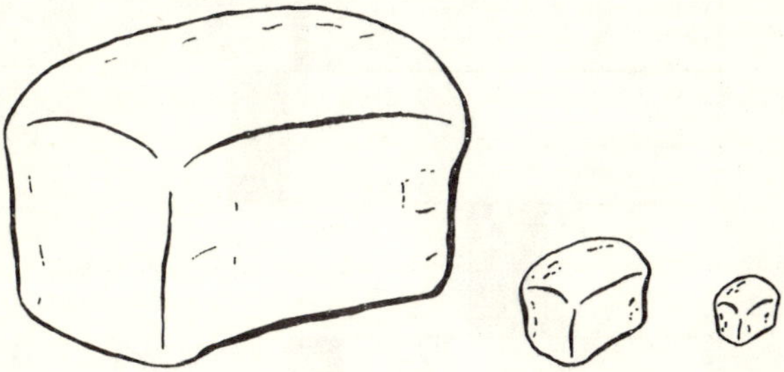

State whether you consider that the diagram gives a clear indication of this relation, giving reasons for your opinion. (London.)

The above example of a *picturegram* or *pictogram* does not give a clear indication of the given relation. It is very misleading because the loaves have been drawn as similar solids with *lengths* in the ratio 7 : 2 : 1. This means that their *volumes* would actually be in the ratio $7^3 : 2^3 : 1^3$, which is 343 : 8 : 1. The eye is better trained to compare lengths than volumes and a comparison of these volumes is even more difficult to assess because they are drawn on a flat sheet of paper. A true representation of the above data would be as shown below.

The popular newspapers and various forms of advertisements provide an infinite variety of pictograms in which colour as well as shape is used with great effect. Pictograms are also called *ideographs*.

141. Bar diagrams. Table 13 A was derived from a survey made by the U.K. Tobacco Retailers' Association. Columns (i) and (ii) give the average annual consumption of cigarettes, in thousands, and pipe tobacco, in kg, by people over 16 years of age. As 1 kg of tobacco is

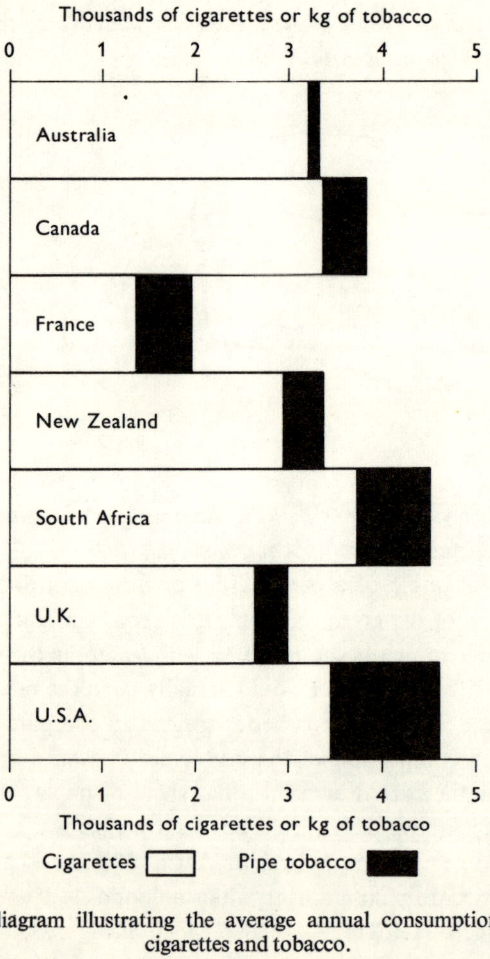

Fig. 36. Bar diagram illustrating the average annual consumption per adult of cigarettes and tobacco.

equivalent to 1000 cigarettes columns (i) and (ii) may be compared without numerical adjustment. Columns (iii), (iv) and (v) are calculated from columns (i) and (ii). Fig. 36 shows the data of columns (i), (ii) and (iii) presented as a *bar diagram*. Fig. 37 is an alternative bar diagram which makes use of the percentages in columns (iv) and (v) to divide the bars lengthwise.

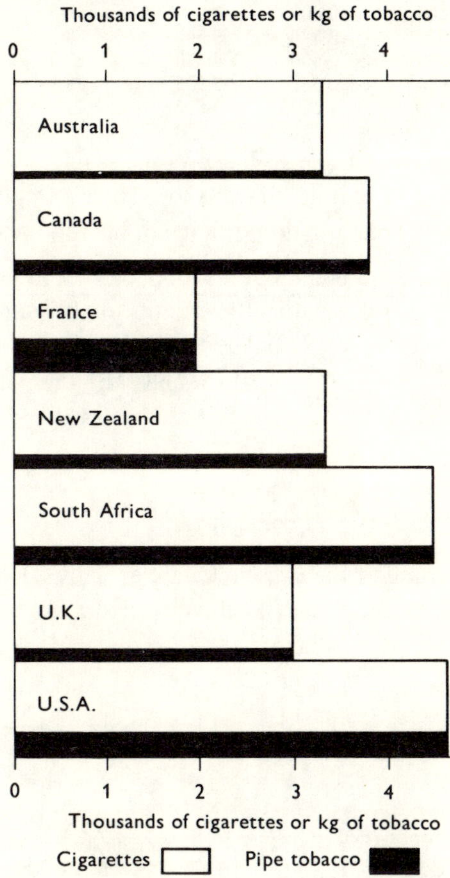

Fig. 37. Alternative bar diagram in which the bars are divided lengthwise according to the percentages of cigarettes and tobacco.

The reader may find it an interesting and artistic exercise to represent the data of columns (i) and (ii) as a picturegram in which a *cigarette* represents 100 cigarettes and a *pipe* represents 0·1 kg of tobacco.

TABLE 13A

Average annual consumption per adult of cigarettes and tobacco

	(i) Thousands of cigarettes	(ii) Weight of pipe tobacco kg	(iii) Total of Columns (i) and (ii)	(iv) Column (i) as percentage of column (iii)	(v) Column (ii) as percentage of column (iii)
Australia	3·220	0·082	3·302	97·5	2·5
Canada	3·350	0·455	3·805	88·0	12·0
France	1·340	0·609	1·949	68·8	31·2
New Zealand	2·970	0·386	3·356	88·5	11·5
South Africa	3·720	0·800	4·520	82·3	17·7
U.K.	2·630	0·355	2·985	88·1	11·9
U.S.A.	3·440	1·182	4·622	74·4	25·6

142. Circular diagrams. Table 13 B shows the savings receipts for the weeks ending 10 and 17 March. Fig. 38 shows these receipts represented by two circles whose areas are proportional to the total receipts. Thus:

$$\frac{\text{The area of the circle representing the receipts for 17 March}}{\text{The area of the circle representing the receipts for 10 March}} = \frac{30{\cdot}230}{24{\cdot}496}.$$

This means that $\dfrac{\pi R^2}{\pi r^2} = \dfrac{30{\cdot}230}{24{\cdot}496}$, where R and r are the radii of the two circles.

Hence

$$\frac{R}{r} = \sqrt{\frac{30{\cdot}230}{24{\cdot}496}} = 1{\cdot}111,$$

and in fig. 38 the radii of the two circles are actually 20 and 22·2 mm.

Further, the first circle is divided into sectors containing angles

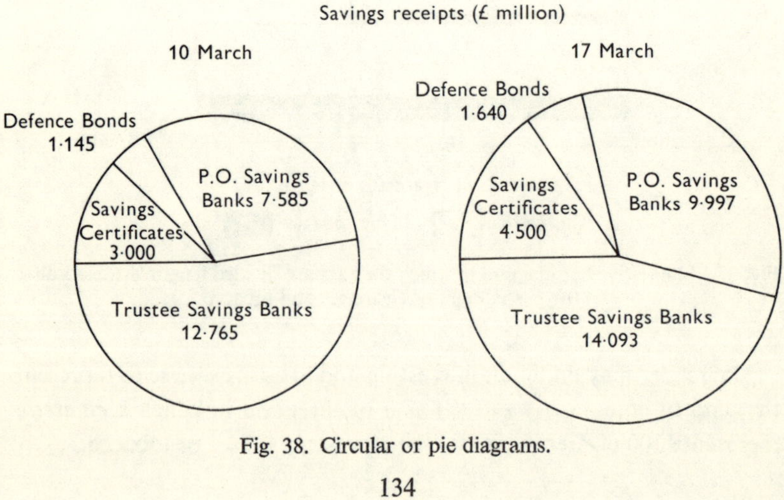

Fig. 38. Circular or pie diagrams.

proportional to 3·000, 1·145, 7·585 and 12·765. This means that the 360 degrees of the circle have to be divided into the ratio

$$3{\cdot}000 : 1{\cdot}145 : 7{\cdot}585 : 12{\cdot}765,$$

and hence the sector representing Savings Certificates contains an angle of $(3{\cdot}000/24{\cdot}496) \times 360$ degrees, that representing Defence Bonds contains an angle of $(1{\cdot}145/24{\cdot}496) \times 360$ degrees and so on. The angles of each sector are shown in table 13 C.

TABLE 13B

Savings receipts (£ million)

	Week ending	
	10 March	17 March
Savings Certificates	3·000	4·500
Defence Bonds	1·145	1·640
P.O. Savings Banks	7·585	9·997
Trustee Savings Banks	12·765	14·093
Total	24·495	30·230

TABLE 13C

The angles of the sectors of the circular diagrams

Savings Certificates	44	53½
Defence Bonds	17	19½
P.O. Savings Banks	111	119
Trustee Savings Banks	118	168
Total	360	360

Circular diagrams such as these are often called, for an obvious reason, *pie diagrams*. They are admirable for comparing totals and for displaying also the separate constituents of the totals but, arithmetically, they involve more difficult calculations than bar diagrams.

143. Exercises.

1. *Average expenditure of industrial and agricultural households in January 1941 (per household)*

	Industrial	Agricultural
	£	£
Food	1·70	1·44
Rent and rates	0·54	0·24
Clothes	0·40	0·26
Fuel and light	0·32	0·25
Other items	1·28	0·73
Total	4·24	2·92

A pictogram is to be made of the above distributions suitable for publication in a periodical, the elements of the pictogram being symbols representing coins. Show, by a rough sketch, how you think the pictogram should be constructed. (Cambridge.)

2. The following figures show the number of boys and girls leaving the Sixth Form Science in a random sample of 130 schools, classified according to their future education or employment.

Leaving for teachers' training colleges	138
Leaving for technical colleges	141
Leaving for universities	635
Leaving to take up paid employment	329
Total	1243

Present the data (i) as a bar diagram, (ii) as a circular diagram.

(Cambridge.)

3. In government and other publications, statistical information is often shown in the form of ideographs; for example, pictures of sacks of different sizes, or alternatively pictures of numbers of sacks of the same size, may be used to show the amounts of wheat and flour obtained from different countries.

Comment on the advantages and disadvantages of these pictorial systems.

(London.)

4. What rules should be observed when diagrams are used to present statistical data?

State which diagrams you would use to illustrate the following kinds of numerical data, giving brief reasons for your choice.

(i) The number of births in a town for each month of the year 1972.

(ii) The total harvest yields of the four main cereal crops, wheat, barley, oats and rye in the United Kingdom for the year 1972.

(iii) The number of pairs of socks of different sizes issued to a unit of soldiers during the year 1972. (London.)

5. Consumers' expenditure on food in the United Kingdom in 1953 is shown below:

	Millions of £
Bread and cereals	527
Meat and bacon	740
Fish	93
Oils and fats	156
Sugar, preserves and confectionery	358
Dairy products	521
Fruit	232
Vegetables	301
Beverages	145
Other manufactured foods	90

Exhibit these facts in a suitable diagram. (London.)

6. The following table gives the production of steel in five districts:

District	*A*	*B*	*C*	*D*	*E*
Production Gg	800	1350	2400	1850	3600

Construct a suitable bar chart and also a pie chart to illustrate these figures. (London.)

7. *University students classified according to sex and living quarters*

	Men	Women	Total
Colleges and hostels	10,800	6,100	16,900
Lodgings	19,000	5,400	24,400
At home	19,900	7,100	27,000

Calculate percentages to compare the distribution of men students among the three types of accommodation with the corresponding distribution of women students, and present them in a self-explanatory table.

Construct a bar diagram to display simultaneously a comparison between the total numbers of students living in each type of accommodation and the sex composition of each total. (J.M.B.)

8. In a certain town each £29 collected from the ratepayers was allocated to the various services in the following way: Education £7·50, Health £7·50, Highways £4·00, Police and Fire Brigade £2·50, Street Lighting £1·50. Other Services £3·50 and Administration £2·50. Construct a diagram which illustrates these data. (J.M.B.)

9. *Customs and Excise, 1955–6* (£ *millions*)

Purchase tax	419
Tobacco	668½
Beer	258
Spirits	125
Light wines	20½
Entertainments	29½
Football pools	20½
Total	1541

Present the above data (i) as a bar diagram, (ii) as a circular diagram.

10. *Japanese shipyards grab the lead*

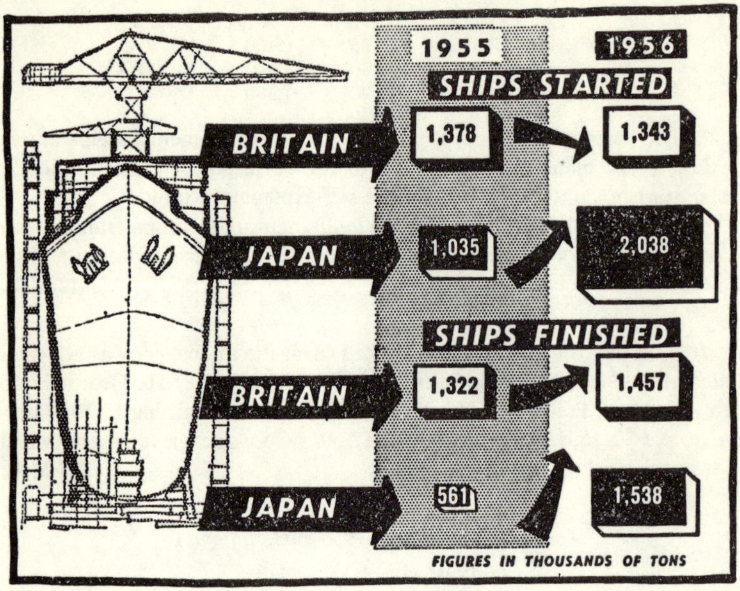

Criticize the above 'Expressograph' (by Michael Rand) which appeared in the *Daily Express*.

Illustrate the data by a simple bar diagram.

Since the number of ships *finished* is of greater importance than the number of ships *started*, do you think that the caption 'Japanese shipyards grab the lead' is fully justified?

144. Sampling. A farmer buying a large consignment of grain might thrust his hand well into the middle of one of the bags and draw out a handful of the contents for careful examination. After repeating this process with three or four more bags selected by chance the farmer's opinion of the whole consignment of grain would be established. This idea of estimating the properties of the whole population by investigating a *sample* taken from the population is widely used in statistical research. If we are considering a *complete* population of observations such as the weights of *all* the men of a certain nationality, the thickness of the cover of a submarine cable at *every* point of its length, the times of *all* journeys by car from A to B, the lengths of *all* components manufactured by a certain machine and we wish to make some inference without looking at every single observation, *we sample.* That is, we consider a set of *typical* observations from which we hope to extract information that in some way is characteristic of the entire population.

145. Bias. The results obtained from a sample should not be applied to the whole population unless the sample is truly *representative* of the whole field of enquiry. Suppose a manufacturer of breakfast foods wishes to test the opinions of the residents of a city about his foods. He might decide to open at random the city telephone directory and question all the subscribers named on that one particular page. If, by chance, the names were to begin with De La Fosse, De La Fongere De La Mare, De La Torre, etc., or perhaps Di Biase, Di Libero, Di Lieto, Di Mambro, etc., or McPherson, Macquire, McRae, etc., or Stein, Steinart, Steinberg, Steiner, etc. the sample would be representative only of a special section of the community. It would be *biased* or *biassed* (both spellings are acceptable). No doubt the reader has already realized also that a sample drawn from the telephone directory is not in any case *representative* because it only considers those members of the population who are telephone subscribers. It is a sample drawn from a particular *stratum* of the population.

146a. Sample designs. In order to avoid bias and to make sure that the entire field of enquiry is fully represented various *sample designs* have been developed. *A sample design is a definite plan, completely determined before any data are collected, for obtaining an unbiased fully representative sample from a given population.*

146b. Cluster sampling. The simple plan of consulting all the subscribers named on one particular page of the telephone directory is an example of *cluster sampling*. A second example would be the estimation of the entire wheat production of a large area from the examination of one particular small square somewhere in the area. A third example would be the establishment of facts about a given region of a city from the questioning of all the families in one block of houses. If, instead of only one, several clusters chosen to a *random* design (see § 146*e*) are used, reliable results can be obtained by this sampling method. The second and third examples described above may also be described as *area sampling* because the clusters are geographic subdivisions.

146c. Systematic sampling. Quite often the simplest way of sampling is to select, say, every 10th card from an index file, every 20th vehicle coming along a road, every 50th component produced by a machine, and so on. Sampling of this sort is called *systematic sampling*. Its great advantage is that it is *spread evenly* over the population. Its one great drawback lies in the possible presence of *hidden periodicities.* The spaces in a car park instantaneously available on the hour (i.e. at 9 a.m., 10 a.m., 11 a.m., etc.) may be quite different from the spaces instantaneously available at, say, 9·25 a.m., 10·25 a.m., 11·25 a.m. etc. Every 50th component produced by a machine may be unsatisfactory if the machine has an irregularity round about every 25th component. The planning of a road may be such that every 10th house may be built to a more imposing design than the rest.

146d. Stratified sampling. To determine public opinion on some political issue it may be a good idea to question separately the age groups *under* 20, 20 *and under* 30, 30 *and under* 50, 50 *and over.* The population in this case has been divided into a number of non overlapping sub populations called *strata.* In *stratified sampling* of this kind we are faced immediately with problems such as: How many strata shall we form? What should the sample size be for each strata? How shall we select the samples in each strata?

A further complication can arise if, for a city, we decide to divide the city into quadrants north-east, north-west, south-east, south-west. Thus we question the north-east under 20 group separately from the north-west under 20 and so on. This process is called *cross-stratification*, and it is quite widely used, particularly in public opinion polls and market surveys.

146e. Random sampling. Suppose the names on a list from which a sample is to be drawn are numbered

1, 2, 3, 4, 5, 6, 7, 8, 9, 10, 11, 12, 13, 14, 15, 16, 17, 18, 19, 20, 21, etc.

and we have 10 cards numbered

1, 2, 3, 4, 5, 6, 7, 8, 9, 0.

Instead of drawing a *systematic sample* as described in § 146*c* by taking every tenth name on the list we can choose a name from the first ten by shuffling the cards and then drawing one at *random.* If we then replace the card, reshuffle and draw, we can choose a name from the second ten. By continuing this process we go through the entire list and give every name on it *an equal chance of selection* thus producing a 10% *random sample.*

A random sample is one for which every member of the population has an equal and independent chance of selection. It is, therefore, fully representative of the whole field of enquiry. Also the Mathematical Theory of Probability can be used in the final interpretation of results obtained by random sampling.

Fortunately the tedious process of shuffling and drawing a card for each group of ten can be avoided by the table of Random Sampling Numbers, table 14 B, which is part of a more extensive table to be found in the *Cambridge Elementary Statistical Tables* by D. V. Lindley and J. C. P. Miller. The digits of table 14 B originally produced by an *Electronic Random Number Indicator* may be regarded as ready-made results of the shuffling and drawing process applied to 10 cards. Moreover, the digits taken in pairs can be used as a substitute for the shuffling and drawing process applied to 100 cards numbered

$$01, \quad 02, \quad \ldots, \quad 99, \quad 00.$$

Thus they may be used to obtain a *1% random sample.*

Our description of random sampling so far has coupled it with *systematic sampling*. It can be coupled equally well with *cluster or area sampling* and with *stratified sampling*. A large wheat producing area can be divided by a *grid map* into, say, 10,000 equal squares, from which a 1% random sample can be drawn; a region of a city may consist of 100 blocks from which a 10% sample can be drawn.

Although the concept of random sampling is easy to understand it often needs an experienced statistician to decide how it can best be applied. If the manufacturer of breakfast goods of § 145 had been determined to get a truly representative random sample he could have drawn a sample of addresses from the Electoral Register. This would have involved a great deal of office work because the Register is kept in the form of a large number of booklets, one for each polling district. When the random sample of addresses had been drawn, interviewers would have had to be sent these addresses. (The sending of interviewers is much more satisfactory than the sending of questionnaires requiring a written postal reply because random sampling demands that every effort be made to obtain the necessary information from the selected addresses.) Also, without doubt the interviewers would have had to make repeated calls at many of the addresses.

The above description shows that, although random sampling is a long expensive operation, it does give a reliable picture of the whole population.

147. Quota sampling. A cheaper and quicker method of sampling than those described in § 146 is *quota sampling*. It is much used in social and market surveys. In this type of sampling each interviewer is given definite instructions about the section of the public he is to question, but the final choice of the actual persons is left to his own convenience and is not predetermined by some carefully operated randomizing plan. Each interviewer might be told, for example, to question 50 people who are to include *at least* 5 men over 40 years of age, 5 men between 20 and 40, 5 men or boys under 20, 5 schoolboys, 5 men dressed in casual clothes, 5 men dressed in formal clothes. Similarly, precise instructions would be given about women and girls. It will be realized, of course, that some of the men in formal clothes will also be in the over 40 class, so that an interviewer might actually cover the first two classes and the last two by interviewing only 10 men for the four classes. The interviewers are stationed throughout the research area at various places in town and country. They make their interviews during the course of a morning, telephone the results to a central office and a final announcement about the survey can be made 24–48 h after its commencement. Quota sampling is relatively inexpensive to operate, and in skilled hands can give good results. It has its pitfalls, however, as the following example shows. A survey was made a year or two ago to find out what percentage of the population took part in gambling. One interviewer reported the astonishing news that all 50 people of his quota were regular gamblers. Subsequent enquiries revealed that he had been interviewing in a street which led to a railway station from which trains to a race meeting were leaving at frequent intervals on that particular morning.

148. The framing of questions and questionnaires. It is very important that questions put to members of the general public by interviewers (or in questionnaires) should be capable of a precise answer. 'Have you had a serious illness recently?' would be unsatisfactory. Instead, the interviewer might ask:

(i) 'Have you been under the doctor during the last twelve months?' The answer 'Yes' being followed by:

(ii) 'How long did the doctor attend you?' and

(iii) 'Did you have to be taken to hospital?'

Again, 'What sort of education have you had?' would be thoroughly unsound. Instead, several questions should be asked such as:

(i) 'What schools did you attend?'

(ii) 'Have you been to a College or University?'

(iii) 'What examinations have you passed?'

In general, questions should be put simply and straightforwardly so that they are easily understood by the unintelligent and uneducated. They should be capable of being answered by 'Yes', 'No', a number, a place, a date or something equally precise. They should be framed tactfully so as to break down any barrier of suspicion or reserve on the part of the informant. (An interviewer is so much more successful than a questionnaire in this respect.) Finally, ambiguity in questions should be avoided at all costs.

148a. Exercises.

1. What is random sampling? Give an example of a sampling scheme which is not random. State the main difference between the two types of sampling you have described. (W.A.E.C.)

2. The table below shows the values, in *millions of Cedi*, of the major exports of Ghana in 1965:

Gold	Cocoa	Manganese	Bauxite	Timber
40	50	25	20	15

Represent this data diagrammatically in two different ways. Comment on the merits of the two representations. (W.A.E.C.)

14

Level of significance

149. The level of significance is a probability. Suppose a man claims that he can toss up any ordinary coin (not two-tailed) and *always* get a *head*. The word *always* presents the difficulty that, although it is easy to prove him wrong, it is impossible to prove him right. Once he obtains a *tail* he is proved wrong. On the other hand, however, we cannot expect him to go on obtaining head after head indefinitely. We must in some way fix a limit after which we accept his claim as justified.

The probabilities of getting 3, 4, 5, 6, 7, heads in succession are respectively $\frac{1}{8}$, $\frac{1}{16}$, $\frac{1}{32}$, $\frac{1}{64}$, $\frac{1}{128}$. Thus, if the man gets 5 heads out of 5, the probability, $\frac{1}{32}$, that he has done so by pure chance is less than $\frac{1}{20}$ or 5% and we say that the man's claim is justified at the 5% *level of significance*. The obtaining of 5 heads out of 5 is a *significant* achievement; it is sound evidence that the man's claim is justified. The probability, $\frac{1}{16}$, of 4 heads out of 4 is greater than 5% and it shows that this achievement is not significant; it could be due to pure chance. On the other hand the probability, $\frac{1}{64}$, is less than $\frac{1}{40}$ or $2\frac{1}{2}$% and it justifies the claim at the $2\frac{1}{2}$% level of significance. The achievement of 6 heads out of 6 is very significant; it is very sound evidence that the claim is justified. Continuing further with the probability, $\frac{1}{128}$, which is less than $\frac{1}{100}$ or 1% we see that 7 heads out of 7 justifies the claim at the 1% level of significance; it is very significant indeed. Finally, if the man gets 10 heads out of 10, the probability, $\frac{1}{1024}$, is less than $\frac{1}{1000}$ or 0·1% and it justifies the claim at the 0·1% level of significance. It is helpful for the beginner to summarize these statements by table 14 A.

TABLE 14A

Levels of significance

Number of heads obtained in succession	Probability	Level of significance	
4 out of 4	$\frac{1}{16}$	greater than 5%	Not significant
5 out of 5	$\frac{1}{32}$	less than 5%	Significant
6 out of 6	$\frac{1}{64}$	less than $2\frac{1}{2}$%	Very significant
7 out of 7	$\frac{1}{128}$	less than 1%	Very significant indeed
10 out of 10	$\frac{1}{1024}$	less than 0·1%	Most significant

It is most important to realize that a claim can seldom, perhaps never, be proved true by statistical calculations. The latter usually justify a claim at a certain level of significance which is a probability. Probabilities in this respect greater than 5 % are generally regarded as not significant and probabilities of 5 % or less as significant.

150. Exercise.

A pack of cards contains the four suits *spades*, *hearts*, *diamonds*, *clubs* in equal numbers. A card is drawn at random. It is then replaced, the cards are shuffled and another is drawn at random. The process is continued. A man claims that he can *always* identify the suit drawn without seeing the face of the card. How many cards must he identify correctly to justify his claim at the 0·1 % level of significance?

151. Example of a significance test on a Binomial population. *Two tennis players A and B are supposed to be equally good but B has won 6 out of the last 8 matches they have played against each other. Does this provide reasonable evidence that A and B are not equally good players?*

The initial assumption that the players are equally good is called the *null hypothesis.* The null hypothesis implies that $p = \frac{1}{2}$ and $q = \frac{1}{2}$ where p and q are the probabilities of success and failure by B in any single match. The null hypothesis leads us to expect that B will win the mean number of matches 4 out of the 8. Assuming that the result of the one match does not affect the next or any other match, that is that the 8 trials are independent,

$$\begin{aligned}(q+p)^8 &= q^8+8q^7p+28q^6p^2+\ldots+28q^2p^6+8qp^7+p^8\\ &= \tfrac{1}{256}+\tfrac{8}{256}+\tfrac{28}{256}+\ldots+\tfrac{28}{256}+\tfrac{8}{256}+\tfrac{1}{256}\end{aligned}$$

indicates that the *total probability of a result as extreme as, or more extreme than,* 6 out of 8 is

$$\begin{aligned}[P(0)+P(1)+P(2)]+[P(6)+P(7)+P(8)] &= \tfrac{74}{256}\\ &= 29\,\%.\end{aligned}$$

This is far greater than 5 % and provides no reason whatever for rejecting the null hypothesis that A and B are equally good. The players may be equally good and yet, by pure chance, B might win *at least* 6 or A win *at least* 6 out of the 8 matches. Note that the phrase 'at least' provides the reason for taking 6, 7, or 8; also one must allow for A having good luck instead of B and so both (0, 1 or 2) and (6, 7 or 8) are taken.

Suppose that B had won 7 matches out of 8. This is not part of the original question but it is an interesting extension of it. The total probability of a result as extreme as, or more extreme than, 7 out of 8 is

$$[P(0)+P(1)]+[P(7)+P(8)] = \tfrac{18}{256}$$
$$= 7\%.$$

This is still not significant at the 5% level.

Only if B wins all 8 matches out of 8 does the total probability $P(0)+P(8) = \tfrac{2}{256}$ fall below 5% and give *sufficient evidence for rejecting the null hypothesis* and in this case, since $\tfrac{2}{256}$ is less than 1%, it is a rejection at the 1% level of significance.

In a significance test of this kind it is important to note that:

(i) We begin by assuming that the players are equally good. This assumption is called the *null hypothesis.* We have to decide finally whether to *accept or reject* the null hypothesis.

(ii) We calculate the total probability of a result as extreme as, or more extreme than the observed result of the trials.

(iii) The calculation (ii) includes the possibility that *A* might have been lucky instead of *B.* It includes *both ends* of the Binomial distribution. It is called a *two-tail* probability.

(iv) If the probability (iii) is less than 5% it is usual to reject the null hypothesis and the observed result is said to be significant at the 5% level.

152. The use of the Normal approximation to the Binomial. In § 90, by means of figs. 13–15, the Normal distribution was shown to be an approximation to the Binomial. The Binomial distribution of the last paragraph has a mean $\mu = 4$ and standard deviation $\sigma = \sqrt{2}$. Now 6, 7 and 8 are 2, 3 and 4 units above the mean and an approximation for $[P(6)+P(7)+P(8)]$ can be obtained from the Normal distribution, $\mu = 4$, $\sigma = \sqrt{2}$ by calculating the probability of any value greater that $1\frac{1}{2}$ units above the mean. As $1\frac{1}{2}$ units above the mean is $(1\frac{1}{2}/\sqrt{2})\,\sigma = 1{\cdot}06\sigma$ the Normal approximation for $[P(6)+P(7)+P(8)]$ is $[1-A(1{\cdot}06)]$.

This cannot be obtained from table 7 B but better tables such as the *Cambridge Elementary Statistical Tables* by D. V. Lindley and J. C. P. Miller show that

$$1-A(1{\cdot}06) = 1-0{\cdot}8554$$
$$= 0{\cdot}1446.$$

Hence the Normal approximation gives

$$[P(0)+P(1)+P(2)]+[P(6)+P(7)+P(8)] = 0{\cdot}2892$$
$$= 29\,\%.$$

This agrees exactly with § 150.

153. A second example of the use of the Normal approximation. *A certain disease which was previously regarded as incurable is now being treated by a new method which, it is claimed, will effect a cure in $\frac{1}{6}$ of all cases treated. In trials of the treatment so far carried out on 120 sufferers from the disease, however, only 12 have been cured. Is this sufficient evidence for rejecting the claim?*

The probability of a cure is $p = \frac{1}{6}$ and when 120 sufferers are treated $(q+p)^{120}$ will give the probabilities $P(0), P(1), \ldots, P(120)$ of the number of cures being 0, 1, ..., 120 respectively. The mean $\mu = 20$ and this is the number of cures we expect out of 120. The standard deviation $\sigma = 10/\sqrt{6}$.

The total probability of the number of cures being as extreme as, or more extreme than, 12 is

$$[P(0)+P(1)+\ldots+P(12)]+[P(28)+P(29)+\ldots+P(120)]$$

and the best way of obtaining this is by the Normal approximation. For this we need the two tails of the Normal distribution outside a range $7\frac{1}{2}$ units above and below the mean.

As $$7\tfrac{1}{2} \text{ units} = 7\tfrac{1}{2} \div (10\sqrt{6}) \text{ standard deviations}$$
$$= 1{\cdot}84 \text{ standard deviations,}$$

the Normal approximation to

$$[P(0)+P(1)+\ldots+P(12)]+[P(28)+P(29)+\ldots+P(120)]$$

is $$2[1-A(1{\cdot}84)] = 2[0{\cdot}0329]$$

by *Cambridge Elementary Statistical Tables*

$$= 0{\cdot}0658$$
$$= 6{\cdot}58\,\%.$$

As this is greater than 5% the 12 cures out of 120 is not sufficient evidence at the 5% level of significance to reject the claim that, in the long run, $\frac{1}{6}$ of all cases will be cures. The situation should be re-examined after more sufferers have been treated.

154. Exercises.

1. It is known that $\frac{1}{3}$ of all bottles made in a certain factory are defective. Calculate the probabilities that a random sample of 10 bottles from the production line contains 0, 1, 2, defective bottles.

A new manufacturing process is introduced and a sample of ten from the product of this new process is found to contain *only one* defective bottle. Would you accept the conclusion that the new process is an improvement on the old? (W.A.E.C.)

2. (*a*) A swimming instructor claims that he can teach 60% of non-swimmers to swim the breadth of the baths in 6 lessons. Assuming this claim to be true and also that the probability of any non-swimmer being successful is independent of the probabilities of the other non-swimmers in the class being successful, calculate for a class of 10 pupils

(i) the probability of more than 8 non-swimmers,

(ii) the probability of less than 2 non-swimmers

learning to swim the breadth in 6 lessons.

(*b*) State briefly the circumstances in which a Normal distribution is a close approximation to a Binomial distribution.

Out of 100 non-swimmers taught by the instructor of (*a*) above only 50 learn to swim the breadth in 6 lessons. Make calculations to decide if this provides sufficient evidence to reject, at the 5% level of significance, the instructor's claim. (Mathematical Association.)

3. A new method of treating rheumatoid-arthritis in children has been developed and its probability of success is claimed to be 70%. ('Success' in this case means that, after treatment, a child is able to lead a full normal life.) Show that if 10 children are treated by this method the chance of more than seven successes is about 38%.

By using the Normal distribution table, or otherwise, show that for a group of 20 children the chance of the number of successes being outside the range 10 to 18 is about 5%. (Mathematical Association.)

4. (i) Given that the probability of an event in a single trial is $\frac{1}{3}$, calculate, to three significant figures, the respective probabilities $P(2)$, $P(4)$, $P(6)$ of 2, 4, 6 occurrences of the event in 18 independent trials.

(ii) Use table 7 B to write down $\frac{1}{2}A(2)$, $\frac{1}{2}A(1)$, $\frac{1}{2}A(0)$ to three significant figures. Explain, fully, why the respective values of (i) are approximately equal to those of (ii). (Mathematical Association.)

5. Over a long period in the past it has been established that 52% of the births in a certain town are males. This year, so far, 60 of the first 100 babies are male. Make calculations to decide if this is sufficient evidence, at the 5% level of significance, to support the view that the percentage of male births has changed.

6. Long experience in the use of a certain rat poison has shown that the percentage kill under natural conditions is 47. Laboratory tests with a new poison gave 27 deaths among 43 rats. Test whether this showed a significant improvement over the standard poison, stating the level of significance.

The new poison was then tested under natural conditions and gave 75 deaths among 148 rats. Test whether the laboratory and field results with this poison are compatible. (Cambridge.)

7. In an investigation of preferences between pre-packed and fresh cut bacon, 69 housewives preferred pre-packed and 52 preferred fresh cut. Does this result provide evidence (at the 5% level of significance) of a difference in acceptability?

Find those values of the percentage of housewives preferring pre-packed which could just be regarded as reasonable, that is which just do not differ from the observed proportion at the 5% level of significance. (Cambridge.)

8. If hens of a certain breed lay eggs on four days a week, on the average, find on how many days during a season of 200 days a poultry keeper with eight hens of this breed will expect to receive at least six eggs.

If in fact he received at least six eggs on 42 days during the season of 200 days, would this evidence suggest that any factor other than chance was operating? (Cambridge.)

9. At a seed-testing station it is found that a proportion 0·4 of a certain type of seed is fertile. By accident the remaining stock of this seed (whose total amount is very large) is completely mixed with an equal quantity of a second type of seed which is believed to be completely infertile. If this latter assumption is true, what is the probability that a seed taken at random from the mixture will germinate?

Each of seven pots is planted with two seeds taken at random from the mixture. Six pots eventually produce one or more plants each. Is this result consistent, at the 5% level of significance, with the infertility postulated for the second type of seed? (Cambridge.)

10. Two brands, A and B, of *fireproof* glass are being tested to decide which one withstands sudden temperature changes best. Use is made of a tub of ice water, C, and an oven, H, as follows:

Both brands are placed in C and then transferred to H which has been set at 150° C.

If neither breaks both are returned to C.

If neither breaks now, both are returned to H which has been advanced to 175° C.

This process continues with 25 degree increases in H until one piece breaks. The piece which breaks first is regarded as poorer in that trial.

If both break together the trial is disregarded.

(i) In 10 trials, brand A broke first 9 times. Is this sufficient evidence to reject, at the 5% level of significance, the assumption that both brands are equally good?

(ii) A report for another set of trials stated that the assumption of equal quality was rejected at the 5% level of significance because brand *A* had always broken first. What is the smallest number of trials that could have been run in this case?

155. Statistics of samples and parameters of populations. The weights of the 32 men *A* and the 256 men *B* given in tables 1 A and 1 B are only of interest in that they enable us to make statements about the weights of men *A* and men *B in general.* They are *random samples* from which we are trying to estimate the properties of the *parent populations* from which they are drawn. To distinguish clearly between the mean and standard deviation of a random sample and the corresponding mean and standard deviation of its parent population Roman small italic letters are used for sample values and the corresponding Greek small letters for the population values. Thus m and s are used to denote the mean and standard deviation of the sample while μ and σ are used to denote the mean and standard deviation of the parent population. The values m and s are often called *statistics,* they are really *secondary statistics* derived from the original observations which are *primary statistics.* The values μ and σ are examples of *parameters.* A parameter is a constant which takes different values for different populations.

In § 157 we shall see that μ usually lies in the range $m \pm 1{\cdot}96\, s/\sqrt{n}$ and nearly always lies in the range $m \pm 3{\cdot}09\, s/\sqrt{n}$ where n is the number of values in the sample.

156. The frequency distribution of means of samples. Table 14 B gives 400 random sampling numbers beginning 5, 5, 9 and continuing to 1, 6, 8. It is an example of 40 samples of 10 taken from a population whose individuals consist of the digits 0, 1, 2, ..., 7, 8, 9 rectangularly distributed. The 40 means of the samples are also given to illustrate the following three important ideas which apply to *samples in general.*

I THE MEAN OF THE MEANS OF SAMPLES
= THE MEAN OF THE INDIVIDUAL OBSERVATIONS.

In this particular case it will be seen that *adding across and dividing by 10* followed by *adding up and dividing by 40* gives the same result as *adding all 400 and dividing by 400.*

Of course, in general, the samples do not cover the whole population but the general rule I applies universally.

TABLE 14B

Random sampling numbers

Each digit is an independent sample from a population in which the digits 0 to 9 each has a probability of $\frac{1}{10}$. The digits are printed in groups of 10 together with the mean of each group

Groups of 10 random numbers	Mean of each group
55 92 52 44 25	*4·3*
60 94 49 57 38	*5·5*
18 48 64 06 57	*4·9*
37 38 37 12 93	*4·6*
36 12 53 96 40	*3·9*
87 85 56 22 37	*5·3*
26 08 51 25 57	*4·1*
37 26 24 23 66	*4·1*
83 99 34 82 37	*5·6*
78 08 92 73 49	*5·7*
90 39 08 21 42	*3·8*
03 27 47 52 03	*3·3*
86 33 76 68 57	*5·9*
80 01 79 33 81	*4·0*
24 85 79 31 96	*5·4*
70 93 14 99 70	*4·9*
34 26 35 91 23	*3·8*
69 03 66 73 82	*5·0*
07 97 85 56 79	*6·3*
75 97 26 90 77	*5·9*
21 04 67 19 13	*3·4*
32 20 99 94 85	*5·1*
20 72 10 95 93	*3·8*
48 96 23 66 45	*5·3*
65 26 11 78 32	*4·1*
86 08 19 91 21	*4·5*
57 06 87 08 15	*4·7*
47 75 62 52 21	*4·1*
20 20 19 26 52	*2·9*
37 14 09 47 16	*4·2*
25 03 61 01 20	*2·0*
98 70 96 36 89	*6·5*
19 17 85 48 51	*4·9*
28 08 70 96 03	*4·3*
84 85 43 95 67	*5·9*
30 27 25 20 88	*3·7*
76 84 41 74 68	*5·5*
69 93 44 05 32	*4·5*
59 56 34 97 07	*5·5*
76 67 90 21 68	*5·2*

Extracted from the *Cambridge Elementary Statistical Tables* by D. V. Lindley and J. C. P. Miller.

II THE STANDARD DEVIATION OF THE MEANS

$$= \frac{\text{THE STANDARD DEVIATION OF THE INDIVIDUALS}}{\sqrt{n}}$$

where n is the number of individuals in each sample (in this case $n = 10$).

It is hoped that the reader will accept this idea without discussion after it is pointed out that the *range* of the means of samples in this particular case is 2·0 to 6·5 compared with the *range* 0 to 9 of the individuals. This is convincing evidence that the *variability* of the means is only a fraction of the *variability* of the individuals but it does not explain the fraction $1/\sqrt{n}$. A full discussion of this is given in my *Statistical Mathematics* (C.U.P. 1973).

III ALTHOUGH THE INDIVIDUALS ARE NOT NORMALLY DISTRIBUTED THE DISTRIBUTION OF THE MEANS TENDS TO NORMAL.

See § 19*b*, Ex. 6.

157. Standard error and confidence limits of large samples. Suppose we have a single sample of n observations (n large). We may regard its mean m and standard deviation s as estimates of μ and σ, the mean and standard deviation of the parent population. Now $\sigma/\sqrt{n}$ is the standard deviation of the means of a large number of samples of n observations similar to our single sample. We can thus regard our single sample as one of many whose means are distributed with standard deviation $s/\sqrt{n}$ about a mean m, s being our estimate of σ. Moreover, it can be shown that even when the distribution of the whole population is far from Normal the distribution of the sample-means will be approximately Normal. Thus, in 95% of cases m will lie in the range given by

$$\mu - 1{\cdot}96\sigma/\sqrt{n} < m < \mu + 1{\cdot}96\sigma/\sqrt{n}.$$

This gives $\mu > m - 1{\cdot}96\sigma/\sqrt{n}$ and $\mu < m + 1{\cdot}96\sigma/\sqrt{n}$.

But we use s as our estimate of σ and hence *it is reasonably certain that the true value of the mean lies between*

$$m - 1{\cdot}96s/\sqrt{n} \quad \text{and} \quad m + 1{\cdot}96s/\sqrt{n}.$$

These limits are known as the 95% *confidence limits* and $s/\sqrt{n}$ is called the *standard error of the mean.*

Further reference to table 7 B shows that 99·8% of the sample-means lie between $m - 3{\cdot}09s/\sqrt{n}$ and $m + 3{\cdot}09s/\sqrt{n}$. These, therefore, are the 99·8% *confidence limits.*

158. Exercises.

1. The mean and the standard deviation of the weights of 11 cartons delivered by machine A, § 43, Ex. 1, were shown in § 51 to be 1·04 kg and 0·03 kg respectively. Show that the standard error of the mean is 0·009 kg and hence calculate (i) the 95% confidence limits, (ii) the 99·8% confidence limits of the mean of the whole output of machine A.

What is the probability of the mean of the whole output being between 1·031 kg and 1·049 kg?

2. The mean and the standard deviation of the 10 measurements of the thickness of the cover of cable A, § 43, Ex. 2, were shown in § 53 to be 7·45 mm and 0·287 mm respectively. Show that the standard error of the mean is 0·091 mm and hence calculate (i) the 95% confidence limits, (ii) the 99·8% confidence limits of the mean thickness of the whole length of the cover.

What is the probability of the mean thickness of the whole length of the cover being between 7·39 mm and 7·51 mm?

159. The significance of a single mean (large samples). In § 157 we saw that, in 95% of cases,

$$m - 1{\cdot}96\sigma/\sqrt{n} < \text{ true } \mu < m + 1{\cdot}96\sigma/\sqrt{n}.$$

If a given value of the mean, μ_0, lies in the above range we can accept it as a possible value of the true μ with 95% confidence; if it does not lie in the range we can reject it with 95% confidence. This statement holds provided

(i) the distribution of μ is Normal, and

(ii) σ is known.

It is approximately true if n is large and s is used for σ. Thus we ask if

$$m - 1{\cdot}96s/\sqrt{n} < \mu_0 < m + 1{\cdot}96s/\sqrt{n}$$

and this is the same as asking if

$$\frac{|\mu_0 - m|}{s/\sqrt{n}} < 1{\cdot}96.$$

Thus *the mean m of a large sample is said to differ significantly from a given value μ_0 if*

$$\frac{|\mu_0 - m|}{s/\sqrt{n}} > 1{\cdot}96.$$

The difference between μ_0 and m in this case is said to be *at the 5% level of significance.*

If
$$\frac{|\mu_0 - m|}{s/\sqrt{n}} > 3{\cdot}09$$

the difference between μ_0 and m is at the 0·2% level of significance; that is to say *very significant indeed* because the probability of the mean of a sample differing from the true mean by an amount large enough to make this possible is $\frac{1}{500}$.

160. Example. *The average breaking strength of steel rods is specified as 20 Mg. The breaking strengths of 100 rods when measured are found to have a mean of 19·9 Mg with a standard deviation of 0·4 Mg. Is a complaint that the rods are not up to specification statistically justified?*

Taking $\mu_0 = 20$, $m = 19{\cdot}9$, $s = 0{\cdot}4$ and $n = 100$,

$$\frac{|\mu_0 - m|}{s/\sqrt{n}} = \frac{0{\cdot}1\sqrt{100}}{0{\cdot}4}$$

$$= 2{\cdot}5.$$

Thus the mean of the sample differs significantly from the specified average and the complaint that the rods are not up to specification is justified.

161. Exercises.

1. A machine producing components to a nominal dimension of 2·000 cm is reset every morning. The first 50 components produced one morning have a mean of 2·001 cm with standard deviation 0·003 cm. Does this provide sufficient evidence that the machine is set too high?

2. Suppose that in the example of § 160 only 50 rods had been tested and that the mean and standard deviation were the same. Would the complaint be justified?

3. Suppose that in the example of § 160 100 rods were tested and that although the mean was the same 19·9 Mg the standard deviation was 0·6 Mg. Would the complaint be justified?

4. A new method is suggested for testing the melting point of metals. The method is applied to a metal whose melting point x° C is known to be 1200° C. For 50 determinations of x, using the new method, the following results were obtained:

$$\Sigma x = 60,800, \quad \Sigma x^2 = 74,214,000.$$

Calculate the mean and standard deviation for the 50 determinations. Would you say that the new method tends to give a value higher than the correct melting point? (W.A.E.C.)

162. The variance of the sums = the sum of the variances. The variance of the differences = the SUM of the variances.

Suppose we have 4 red cards numbered 6, 7, 9, 10 and 3 black cards numbered 1, 3, 5. The red and black cards represent two independent populations. If we select at random 1 red card and 1 black the 12 possible pairs of numbers we might obtain are:

6 and 1, 6 and 3, 6 and 5, 7 and 1, 7 and 3, 7 and 5, 9 and 1, 9 and 3, 9 and 5, 10 and 1, 10 and 3, 10 and 5.

Now the mean and variance of the 4 *red numbers* are 8 and 10/4 respectively and the mean and variance of the 3 *black numbers* are 3 and 8/3 respectively. Further, the mean of the sums of the pairs of numbers is

$$\tfrac{1}{12}\{7+9+11+8+10+12+10+12+14+11+13+15\} = 11$$

and the variance is

$$\tfrac{1}{12}\{16+4+0+9+1+1+1+1+9+0+4+16\} = \tfrac{62}{12}.$$

Thus *the mean of the sums = the sum of the means*

and *the variance of the sums = the sum of the variances.*

Also, the mean of the differences of the pairs of numbers is

$$\tfrac{1}{12}\{5+3+1+6+4+2+8+6+4+9+7+5\} = 5$$

and the variance is

$$\tfrac{1}{12}\{0+4+16+1+1+9+9+1+1+16+4+0\} = \tfrac{62}{12}.$$

Thus *the mean of the differences = the difference of the means*
and *the variance of the differences = the SUM of the variances.*

The above special cases are examples of two important general theorems which may be stated as follows:

If the mean and variance of the m values $x_1, x_2, x_3, \ldots, x_m$ are $\bar{x}$ and s_x^2 and the mean and variance of the n values $y_1, y_2, y_3, \ldots, y_n$ are $\bar{y}$ and s_y^2 then (i) *the mean and variance of the mn values of z given by $z = x+y$ are $\bar{x}+\bar{y}$ and $s_x^2+s_y^2$, and* (ii) *the mean and variance of the mn values of z given by $z = x-y$ are $\bar{x}-\bar{y}$ and $s_x^2+{}_y^2s$.* (It being understood that x and y represent two independent populations.)

163. Experiment with two coloured dice. If a single die is thrown a large number of times the scores 1, 2, 3, 4, 5, 6 are all equally probable and the mean score will therefore approximate to

$$\tfrac{1}{6}[1+2+3+4+5+6] = 3\tfrac{1}{2}. \tag{1}$$

While the variance of the scores will approximate to

$$\tfrac{1}{6}(6\tfrac{1}{4}+2\tfrac{1}{4}+\tfrac{1}{4}+\tfrac{1}{4}+2\tfrac{1}{4}+6\tfrac{1}{4}) = 2\tfrac{11}{12}. \tag{2}$$

If a white die and a black die are thrown together a large number of times and a record is kept of

(*a*) the white score + the black score for each throw,

(*b*) the white score − the black score for each throw,

§ 162 leads us to expect:

$$\text{the mean of the sums } (a) = 7, \tag{3}$$

$$\text{the variance of the sums } (a) = 5\tfrac{5}{6}, \tag{4}$$

$$\text{the mean of the differences } (b) = 0, \tag{5}$$

$$\text{the variance of the differences } (b) = 5\tfrac{5}{6}. \tag{6}$$

The facts of (1), (2), (3), (4), (5) and (6) above should be tested experimentally by throwing a single die

(i) 10 times, (ii) 50 times, (iii) 100 times,

and by throwing a pair of differently coloured dice

(i) 10 times, (ii) 50 times, (iii) 100 times.

164. Three or more dice. If the ideas of §§ 162 and 163 are extended to the mean and variance of the total score when three or more dice are thrown together the results shown in table 14 C hold:

TABLE 14C

No. of dice thrown together	Mean score	Variance of scores	Standard deviation of scores
1	$3\frac{1}{2}$	$2\frac{11}{12}$	1·708
2	7	$5\frac{5}{6}$	2·416
3	$10\frac{1}{2}$	$8\frac{3}{4}$	2·958
4	14	$11\frac{2}{3}$	3·416
5	$17\frac{1}{2}$	$14\frac{7}{12}$	3·819

An important generalization of the above is:

If $x_1, x_2, x_3, \ldots, x_n$ are n statistically independent variables

(i) *the mean of* $(x_1+x_2+x_3+\ldots+x_n)$ = *mean of* x_1 + *mean of* x_2 + *mean of* x_3 + … + *mean of* x_n,

(ii) *the variance of* $(x_1+x_2+x_3+\ldots+x_n)$ = *variance of* x_1 + *variance of* x_2 + *variance of* x_3 + … + *variance of* x_n.

A practical application follows.

165. An air travel example. *The following table expresses in statistical terms a summary of a statement made by a man who travels regularly by air from a city centre A to a city centre B:*

	Mean time (minutes)	Standard deviation (minutes)
Road journey from city centre *A* to airport *A*	30	5
Wait at airport *A* before take-off	40	7
Flight from airport *A* to airport *B*	50	4
Wait at airport *B*	20	3
Road journey from airport *B* to city centre *B*	12	2

Assuming that all the times are Normally and independently distributed show that on 95% of occasions the time of the journey from city centre A to city centre B is likely to be within the range 132 to 172 minutes.

The mean of the *sum* of the times = the sum of the mean times
= 152 minutes

and the variance of the *sum* $= 5^2+7^2+4^2+3^2+2^2$
$= 103 \text{ min}^2$.

Thus the standard deviation of the *sum* of the times is $\sqrt{103} = 10{\cdot}15$ minutes.

Moreover, if the separate times are Normally and independently distributed it can be proved that their *sum* is Normally distributed. Hence, by table 7 B, on 95% of occasions the time of the whole journey will be in the range $152 \pm 1{\cdot}96 \times 10{\cdot}15$ minutes; that is, between 132 and 172 minutes.

166. Exercises.

1. A certain firm mass-produced machines which, in the course of their assembly passed through four workshops, *A*, *B*, *C* and *D*. A record of the times taken in each workshop and the times of transit from one workshop to the next was kept and from it the following summary showing the means and standard deviations of the times was published:

	Mean time (h)	Standard deviation (h)
Workshop *A*	3·48	0·25
Transit from *A* to *B*	0·23	0·05
Workshop *B*	4·56	0·30
Transit from *B* to *C*	0·53	0·12
Workshop *C*	1·91	0·20
Transit from *C* to *D*	0·32	0·10
Workshop *D*	2·67	0·20

Assuming that the workshop and transit times were independently Normally distributed, calculate the mean and standard deviation of the times for the complete assembly of the machines, and show that only 1% of the machines were assembled in less than $12\frac{1}{2}$ h while 6% took over $14\frac{1}{2}$ h. (J.M.B.)

2. Four athletes specialize in running 200 m, 200 m, 400 m and 800 m respectively. They train as a team for a 1600 m medley relay race. During training their mean times for their respective distances are 23·9 sec, 24·1 sec, 53.6 sec and 2 min 7·4 sec, and the corresponding standard deviations are 0·3 sec, 0·3 sec, 0·8 sec and 1·8 sec. Assuming that their individual times are Normally and independently distributed, estimate the mean and standard deviation of the times in which the team covers 1600 m. Also estimate the probability that, on any particular occasion, the time will be 3 min 45 sec or less. (J.M.B.)

3. At the assembly stage in the manufacture of a knitting machine, four sections of bakelite of lengths x_1, x_2, x_3, x_4 are drawn at random from a box containing a large number of these sections and fitted between two pieces of metal on the partially assembled machine, these being separated by a distance y as shown in the diagram. The fit is considered satisfactory if the clearance $y-x_1-x_2-x_3-x_4$ lies between zero and 0·10 cm (see fig. 39).

If the lengths of the pieces in the box vary randomly about a mean of 4·00 cm, with a standard deviation of 0·015 cm, and if y varies randomly from machine to machine with mean 16·06 cm and standard deviation 0·024 cm, find the mean and standard deviation of the distribution of $y-x_1-x_2-x_3-x_4$.

Hence determine the proportion of assemblies giving a satisfactory fit at the first attempt. (J.M.B.)

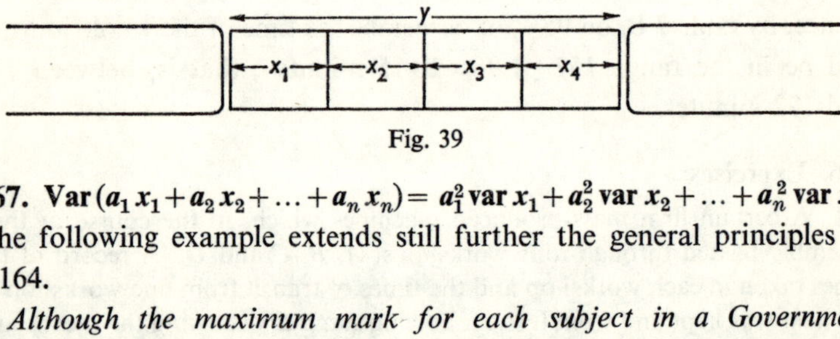

Fig. 39

167. Var $(a_1 x_1 + a_2 x_2 + \ldots + a_n x_n) = a_1^2 \text{ var } x_1 + a_2^2 \text{ var } x_2 + \ldots + a_n^2 \text{ var } x_n$. The following example extends still further the general principles of § 164.

Although the maximum mark for each subject in a Government Servants' Promotion Examination is 100 some of the subjects are regarded as more important than others. The subjects, therefore, are WEIGHTED according to the scheme shown in the table below and the maximum total mark for the five subjects is 1000. Use the given data to obtain the mean total mark and its standard deviation.

	Weight	Mean	Standard deviation
Proficiency in reading and writing	3	55	10
Proficiency in Computation	3	63	9
Government law	2	69	8
History	1	51	12
Geography	1	58	11

If the marks of any individual candidate are x_1, x_2, x_3, x_4, x_5, their *weighted* total is

$$3x_1+3x_2+2x_3+x_4+x_5.$$

Now it can be proved mathematically that if x_1, x_2, x_3, x_4, x_5 are *independent* variables and a_1, a_2, a_3, a_4, a_5 are constants then

(i) the mean of $(a_1 x_1+a_2 x_2+\ldots+a_5 x_5)$
$$= a_1(\text{mean of } x_1)+a_2(\text{mean of } x_2)+\ldots+a_5(\text{mean of } x_5),$$

(ii) the variance of $(a_1 x_1+a_2 x_2+\ldots+a_5 x_5)$
$$= a_1^2(\text{var } x_1)+a_2^2(\text{var } x_2)+\ldots+a_5^2(\text{var } x_5).$$

Thus in the example under consideration, assuming that the separate marks are independent, the mean total mark is

$$3\times 55+3\times 63+2\times 69+1\times 51+1\times 58 = 601$$

and the variance of the totals is

$$3^2\times 10^2+3^2\times 9^2+2^2\times 8^2+1^2\times 12^2+1^2\times 11^2 = 2150.$$

The standard deviation of the totals is, therefore,

$$\sqrt{2150} = 46{\cdot}4.$$

If it is also known that the marks in the separate subjects are *Normally distributed* then the totals will be Normally distributed about the mean 601 with standard deviation 46·4.

168. Example. *A large consignment of mercury is supplied in small containers. The volumes of mercury in the containers are distributed about a mean of 250 c.c. with a standard deviation of 2 c.c. and the weights of the empty containers are distributed about a mean of 456 g with a standard deviation of 7 g. Calculate the mean and standard deviation of the full containers given that the specific gravity of mercury is 13·6.*

Taking the mean of $x_1 = 250$, the mean of $x_2 = 456$,

the variance of $x_1 = 4$, the variance of $x_2 = 49$,

$$a_1 = 13{\cdot}6, \quad a_2 = 1$$

we obtain (i) the mean weight of the full containers is

$$13{\cdot}6\times 250+1\times 456 = 3856 \text{ g}$$

and (ii) the variance of these weights is

$$13{\cdot}6^2\times 4+1^2\times 49 = 789 \text{ g}^2.$$

Thus the standard deviation is 28·1 g.

169. The significance of the difference between the means of two large samples. Consider the two large samples details of which are given in table 14 D.

TABLE 14D

	Sample 1	Sample 2
No. of observations	n_1	n_2
Mean	m_1	m_2
Standard deviation	s_1	s_2
Standard error	$s_1/\sqrt{n_1}$	$s_2/\sqrt{n_2}$

As explained in § 157, s_1^2/n_1 is an estimate of the variance of the means m_1 of many samples of n_1 observations drawn from the same population as sample 1 and s_2^2/n_2 is an estimate of the variance of the means m_2 of many samples of n_2 observations drawn from the same population as sample 2. Hence the estimated variance of the differences (m_1-m_2) is $\dfrac{s_1^2}{n_1}+\dfrac{s_2^2}{n_2}$. Thus, *the standard error of the differences is*

$$\sqrt{\left(\frac{s_1^2}{n_1}+\frac{s_2^2}{n_2}\right)}.$$

Moreover, the mean of the differences being equal to the difference of the means is (m_1-m_2) and if sample 1 and sample 2 both belong to the same population, the mean of the differences should not differ significantly from zero. Hence, for large samples, if

$$\frac{|m_1-m_2|}{\sqrt{\left(\dfrac{s_1^2}{n_1}+\dfrac{s_2^2}{n_2}\right)}}>1{\cdot}96,$$

the difference between the means is significant at the 5% level. This means that the chance of such a large 'difference' when the samples are from the same parent population is less than $\frac{1}{20}$.

If

$$\frac{|m_1-m_2|}{\sqrt{\left(\dfrac{s_1^2}{n_1}+\dfrac{s_2^2}{n_2}\right)}}>3{\cdot}09,$$

the difference between the means is significant at the 0·2% level and the chance of such a large 'difference' when the samples are from the same parent population is $\frac{1}{500}$.

170. A comparison of two methods of slimming. *When 144 girls volunteered to undergo slimming treatment they were divided into two separate groups A and B each of 72 girls. The two groups were then submitted to two different methods of slimming for 3 months. A comparison of the two treatments is summarized in the following table:*

Weight in kg

	Before		After	
	Mean	Standard deviation	Mean	Standard deviation
Group A	60	10	53	8
Group B	57	10	50	8

Discuss the differences between the means statistically.

(i) Comparing groups A and B before the treatment we see that

$$\frac{|m_1-m_2|}{\sqrt{\left(\frac{s_1^2}{n_1}+\frac{s_2^2}{n_2}\right)}} = \frac{3}{\sqrt{\left(\frac{100}{72}+\frac{100}{72}\right)}} = 1{\cdot}8.$$

Thus the difference between the means is not significant at the 5% level.

(ii) Comparing groups A and B after the treatment

$$\frac{|m_1-m_2|}{\sqrt{\left(\frac{n_1^2}{s_1}+\frac{n_2^2}{s_2}\right)}} = \frac{3}{\sqrt{\left(\frac{64}{72}+\frac{64}{72}\right)}} = 2{\cdot}25.$$

The difference between the means is then significant at the 5% level. Although the actual difference is the same the variability of weight is reduced to such an extent that the difference of 3 is no longer negligible.

(iii) For group A only, comparing the mean before treatment with the mean after treatment we see that

$$\frac{|m_1-m_2|}{\sqrt{\left(\frac{s_1^2}{n_1}+\frac{s_2^2}{n_2}\right)}} = \frac{7}{\sqrt{\left(\frac{100}{72}+\frac{64}{72}\right)}} = 4{\cdot}6.$$

Thus the difference between the means is significant at the 0·2% level, that is to say very significant indeed.

(iv) Similarly, for group B only, the difference between the means before and after treatment is significant at the 0·2% level.

Summarizing finally we may state that in the beginning, the weights of the girls in group A could not be regarded as different from the weights of the girls in group *B*. Both treatments are proved very effective, treatment *B* being probably more effective than treatment *A*.

171. Exercises.

1. In order to find out whether the average speed of motor-vehicles leaving London is different from that of motor-vehicles entering London, cars and motor-cycles were timed over a stretch of the Portsmouth road. The following table shows the results of the investigation:

	Leaving London	Entering London
No. of vehicles timed	50	50
Mean time in sec	17·04	18·38
Variance in sec^2	8·846	9·106

Determine whether the difference between the means is significant (i) at the 5% level, (ii) at the 1% level. Comment on the meaning of your result. (J.M.B.)

2. A firm which manufactures lead-covered submarine cable suspected that the lead was being put on more thickly by its night workers than by its day workers. To keep down the cost the lead must be as thin as possible but must be nowhere less than 5 mm thick if it is to withstand the action of the sea water and general wear and tear. Part of an investigation carried out is summarized in the following table:

	Day work	Night work
No. of places at which the thickness of the lead cover was measured	100	100
Mean thickness (mm)	7·300	7·450
Standard deviation (mm)	0·525	0·475

Determine whether the difference between the means is significant or not and make comments on the implications of the given figures and your result. (J.M.B)

3. *A test on the length of life of two types of car tyres*

	Type *A*	Type *B*
Number of tyres tested	50	40
Mean life in miles	24,000	26,000
Standard deviation (miles)	2,500	3,000

Show that the difference between the means is significant at the 0·2% level.

15

Miscellaneous exercises

1. Describe briefly a statistical investigation in which you have taken part, referring particularly to

(*a*) preliminary planning,
(*b*) the methods used to collect and tabulate data,
(*c*) the methods used to analyse the data, and
(*d*) any conclusions reached. (London.)

2. The following table gives details of weekly income and of weekly expenditure on food of a group of four families:

Family	Total family income (£)	Expenditure per head on food (£)	Total number in family
A	14·8	1·6	4
B	28·2	1·8	6
C	16·4	1·4	5
D	35·2	2·1	4

Find, in £ to 2 decimals, the average income per head of the group and the average expenditure on food per family.

Which families spend (*a*) more, (*b*) less than the average on food and by how much? (London.)

3. The numbers of thousands of live births registered in three regions *A*, *B* and *C* in a certain year were:

	Region *A*	Region *B*	Region *C*	Total
1st quarter	10·4	23·5	7·3	—
2nd quarter	10·1	24·1	7·7	—
3rd quarter	10·0	22·8	7·1	—
4th quarter	9·4	21·9	6·8	—

Find for each quarter the total of the numbers registered.

Construct a bar diagram to illustrate simultaneously these totals and the contributions to these totals from each region. (London.)

4. Explain carefully the meaning of the following terms: (*a*) index number, (*b*) frequency distribution, (*c*) median, (*d*) mode, and (*e*) dispersion. (London.)

5. The following table shows the monthly production of cane sugar in units of 100 tons for the period July 1953 to June 1955:

	J	F	M	A	M	J	J	A	S	O	N	D
1953	—	—	—	—	—	—	22	27	24	21	18	16
1954	17	17	16	19	23	24	26	23	26	19	20	15
1955	17	19	19	25	27	28	—	—	—	—	—	—

Construct a table showing the twelve-monthly moving average correct to the nearest 10 tons and draw a graph to show the trend. (London.)

6. In a botanical experiment the lengths of 100 leaves from a certain species of plant were measured and the frequencies tabulated as follows:

Length of leaf (mm)	Frequency
20–24	1
25–29	5
30–34	10
35–39	19
40–44	25
45–49	21
50–54	15
55–59	3
60–64	1

Construct a histogram to show the frequency distribution and state the mode. (London.)

7. Using a frequency distribution of Question 6, with a working mean of 42, calculate the mean length of leaf and the standard deviation from the mean. (London.)

8. From the data of Question 6 construct a cumulative frequency table and draw the ogive. Use the ogive to determine (i) the median length of leaf, and (ii) the difference in length between the first and third quartiles. (London.)

9. The percentage marks for ten candidates for Intelligence and Arithmetic tests were as follows:

Candidate	*A*	*B*	*C*	*D*	*E*	*F*	*G*	*H*	*J*	*K*
Intelligence test	27	39	48	42	47	55	35	30	32	45
Arithmetic test	40	44	65	67	80	84	60	57	46	55

Plot a scatter diagram and a line of best fit.

A candidate X scored 40 in the Intelligence test and a candidate Y scored 55 in the Arithmetic test. Each was absent from the other test. Use your diagram to estimate a comparison between the performances of X and Y. (London.)

10. Arrange the table of results in Question 9 according to ranks and calculate the rank correlation coefficient.

State what significance, if any, you would attach to the result. (London.)

11. Give an example to illustrate the construction and use of each of the following: (i) histogram or column diagram, (ii) pie chart, (iii) cumulative frequency diagram. (London.)

12. In a certain region the total area sown with wheat was 10,324 km^2, *to the nearest km^2*. The average yield was 280,000 kg per *km^2*, *correct to the nearest 1000 kg*. Estimate the total yield of wheat in kg as accurately as the data allow.

Calculate also, in kg, the extreme limits between which the total yield must lie. (London.)

13. Twenty boys took two papers in a Mathematics examination and their percentage marks were as follows:

Paper I	51	42	42	54	39	27	72	57	42	42
Paper II	60	51	45	60	30	30	57	57	39	54
Paper I	36	33	42	75	39	57	78	42	42	60
Paper II	39	21	36	72	45	54	78	42	48	63

Plot a scatter diagram of the marks in paper I against the marks in paper II and draw a line of regression.

Explain how this line will indicate if the candidates found one paper more difficult than the other.

State two other ways of comparing the difficulty of the two papers. (J.M.B.)

14. The following table shows the coal production (in millions of Mg.) of (*a*) United States of America, (*b*) Great Britain, and (*c*) Germany, for the first half of the twentieth century:

Years	(*a*)	(*b*)	(*c*)
1900–04	300	200	190
1905–09	400	250	200
1910–14	450	300	250
1915–19	540	250	220
1920–24	530	230	230
1925–29	540	220	300
1930–34	400	210	270
1935–39	400	230	350
1940–44	550	200	440
1945–49	550	200	230

Construct a histogram to illustrate both the relative and the combined production of the three countries. (London.)

15. The following pairs of values of x and y were obtained by experiment:

x	0·5	1·1	1·8	2·2	2·7	3·2	3·8	4·7
y	2·3	2·5	2·9	3·2	3·4	3·5	4·0	4·3

Plot these results in a diagram and draw a line of best fit. Use your graph to estimate (i) the value of y when $x = 2·5$, and (ii) the value of x when $y = 5·0$. (London.)

16. The index numbers of commodities A, B and C in 1956, taking 1939 as base are shown below:

	1939	1956
A	100	186
B	100	242
C	100	282

Calculate the weighted average of these index numbers for 1956 when A, B and C are given weights of 4, 2 and 3 respectively.

Calculate, each to the nearest whole number, the 1939 index numbers referred to 1956 as base. (London.)

17. The following table was drawn up from the height measurements of 100 recruits:

Height (mm) (mid-point of interval)	1600	1650	1700	1750	1800	1850
Frequency	3	12	32	24	21	8

Construct a cumulative frequency table and draw the ogive. Use the ogive to determine the median height.

What is the modal height of the distribution? (London.)

18. Using the frequency distribution of Question 17, with a working mean of 1700 mm, find the mean height and the standard deviation from the mean. (London.)

19. The following table shows estimated personal expenditure (in £1,000,000) in Great Britain on fuel and light:

	Quarter			
	(1)	(2)	(3)	(4)
1953	133	106	92	116
1954	144	116	100	125
1955	155	121	106	139
1956	175	—	—	—

Calculate the four-quarterly moving average and draw a graph to show the trend. (London.)

20. The marks of 10 students in Arithmetic and Algebra tests were:

Student	*A*	*B*	*C*	*D*	*E*	*F*	*G*	*H*	*I*	*J*
Arithmetic	49	60	41	34	21	42	43	65	45	63
Algebra	39	74	33	32	13	31	34	71	57	40

Calculate the coefficient of rank correlation between the performances in Arithmetic and Algebra. (London.)

21. *Weight of litter left by visitors in London parks after a holiday*

	Left in baskets (kg)	Left on ground (kg)
Hyde Park	3050	1050
Kensington Gardens	800	350
St James's and the Green Parks	1200	300
Greenwich Park	350	250
Regent's Park	3300	1550
Richmond Park	400	350
Hampton Court	700	400
Bushey Park	200	250
Total	10,000	4500

Display the above data by two circular diagrams, the total weight of litter left in baskets being represented by a circle of radius 45 mm. All the necessary calculations should be shown clearly and the radius of the circle representing the total weight of litter left on the ground should be stated. (J.M.B.)

22. (*a*) Four numbers a, b, c, d are such that b is 90% of a, c is 80% of b and d is 120% of c.

Express (i) c as a percentage of a, (ii) d as a percentage of a.

(*b*) A form consists of n_1 boys and n_2 girls. In an examination the average mark of the boys was x and the average mark of the girls was y. Find the average mark of the whole form. (J.M.B.)

23. *Weights (kg) of crews taking part in the finals at the Henley Royal Regatta 1956*

	1	2	3	4	5	6	7	8	Cox
Royal Air Force	73	78	80	84	75	76	75	71	63
Princeton University, U.S.A.	70	70	72	78	79	73	69	69	57
Peterhouse, Cambridge	79	78	80	76	85	81	73	78	57
Magdalene College, Cambridge	80	78	75	83	78	77	70	70	53
St Paul's School	74	70	78	72	85	82	74	75	51
Eton College	70	77	80	77	86	80	76	85	53
L'Armé Francaise	79	77	80	84	103	86	84	76	51
Roddklubben, Sweden	78	85	75	82	87	86	82	79	55

Form a frequency distribution for the above weights of 72 men and construct the histogram. The following grouping is appropriate: $49\frac{1}{2}$–$54\frac{1}{2}$, $54\frac{1}{2}$–$59\frac{1}{2}$, $59\frac{1}{2}$–$64\frac{1}{2}$, etc. Comment on any special features of the histogram. (J.M.B.)

(Note: In racing rowing boats there are eight oarsmen and a cox. The cox steers the boat and is usually lighter than the oarsmen.)

24. The first three rows of the following table show extracts from the vital statistics of a certain town while the last row gives an assumed standard distribution per thousand:

Age group	0–9	10–39	40–59	60–
Population in thousands	16	48	12	4
Deaths in 1955	212	247	291	337
Standard age distribution	150	400	300	150

Calculate the crude and standardized death-rates and explain why crude death-rates are considered inadequate. (J.M.B.)

25. Describe any statistical investigation in which you have assisted (or would like to assist). The description should state clearly

(i) the object of the investigation,

(ii) the methods of obtaining information,

(iii) the precautions necessary for avoiding bias,

(iv) the method of tabulating the results. (J.M.B.)

26. The following table shows the frequency distribution of the examination marks of 700 candidates in a Mathematics examination:

Examination marks	Number of candidates	Examination marks	Number of candidates
10–19	10	60–69	148
20–29	32	70–79	127
30–39	45	80–89	90
40–49	87	90–99	44
50–59	117		

Taking a working zero in the class 60–69 and a unit of 10 marks calculate the arithmetic mean.

Construct a cumulative frequency curve and use it to determine:

(i) the percentage of candidates passing if the pass mark is 55;

(ii) the lowest mark for distinction if 5% of the candidates are to be given distinction. (J.M.B.)

27. *Analysis by size of the cinemas in a certain region*

Seating capacity	Number of cinemas
Under 251	3
251– 500	70
501– 750	127
751–1000	125
1001–1250	56
1251–1500	25
1501–1750	17
1751–2000	7
2001–2250	5
2251–2500	4
2501–2750	3
2751–3000	2
Total	444

Taking a working zero in the class 751–1000 and a unit of 250 seats, calculate the arithmetic mean seating capacity of the cinemas.

Construct a cumulative frequency curve and use it to determine the median. (J.M.B.)

28. (*a*) The percentage of teachers who are men is a, the percentage of teachers who have honours degrees is b, and the percentage of teachers with honours degrees who are men is c. Deduce the percentage of men teachers who have honours degrees.

(*b*) In a survey designed to ascertain the views of grammar school pupils the following questions are put to samples of boys from grammar schools, the samples being selected by the headmasters:

(i) *Do you think that the more modern subjects, such as Physics, Chemistry, and Biology, are more important than the traditional subjects, such as History, Latin and Greek?*

(ii) *Do you intend to stay on at school after the minimum leaving age permitted by law?*

Criticize the questions and the method of selecting the sample from the point of view of potential bias. (J.M.B.)

29. The following, in chronological order, were the annual numbers of injuries (in units of a thousand) suffered by mine workers in a certain county in 28 successive years: 195, 178, 91, 173, 162, 176, 166, 141, 126, 122, 133, 134, 136, 141, 132, 134, 146, 158, 167, 174, 177, 181, 167, 163, 183, 229, 238, 234.

Taking class intervals 90–, 120–, 150–, 180–, and 210–240, construct a frequency table for the data, and draw a histogram. Point out its most prominent features.

Discuss briefly whether the histogram alone is a fair summary of the data, and whether it would be a fair conclusion that in the first 24 years the mines were rarely as dangerous as the were during the last 4 years. (J.M.B.)

30. *Frequency of instantaneous speeds of motor vehicles at a point in a busy street*

Speed (km/h)	Percentage frequency	Speed (km/h)	Percentage frequency
0–	0	32–	25·5
8–	5·5	40–	9·0
16–	15·0	48–	3·0
24–	41·5	56–64	0·5

Taking a working zero in the class 24– and a unit of 8 km/h, calculate the arithmetic mean speed, using the frequency table above. Construct a cumulative percentage frequency graph, and use it to estimate the 10–90 percentile range. (J.M.B.)

31. *U.K. index of retail prices, 17 September 1968*

	Price index (16 Jan. 1962 = 100)	1968 Weight
Food	122·6	263
Alcoholic drink	127·2	63
Tobacco	127·8	66
Housing	142·2	121
Fuel and light	133·2	62
Durable household goods	114·1	59
Clothing and footwear	114·1	89
Transport and vehicles	121·0	120
Services	133·7	56
Miscellaneous goods	127·3	60
Meals bought and consumed outside the home	129·4	41

(Source: *Annual Abstract of Statistics, 1968.*)

Using the given weights, calculate the weighted arithmetic mean of the above price indices.

32. *U.K. index of retail prices, 16 September 1969*

	Price index (16 Jan. 1962 = 100)	1969 Weight
Food	131·3	254
Alcoholic drink	136·2	64
Tobacco	135·8	68
Housing	147·6	118
Fuel and light	135·4	61
Durable household goods	119·0	60
Clothing and footwear	118·8	86
Transport and vehicles	124·3	124
Services	133·1	66
Miscellaneous goods	143·3	57
Meals bought and consumed outside the house	137·2	42

(Source: *Annual Abstract of Statistics, 1969.*)

Using the given weights, calculate the weighted arithmetic mean of the above price indices.

33. *U.K. index of retail price of food, 17 November 1970*

	Price index (16 Jan. 62 = 100)	1970 Weight
Bread, flour, cereals, biscuits and cakes	149	37
Meat and bacon	153	65
Fish	154	9
Butter, margarine, lard and cooking fat	125	11
Milk, cheese and eggs	137	38
Tea, coffee, cocoa, soft drinks	122	15
Sugar, preserves and confectionery	157	24
Vegetables: fresh, dried and canned	140	27
Fruit: fresh, dried, canned	112	13
Other food	142	16
Total		255

(Source: *Monthly Digest of Statistics*, Dec. 1970.)

By working with an arbitrary origin of 140, or otherwise, obtain the 'All items' index of retail price of food for 17 Nov. 1970 from the above price indices.

Give a brief account of (i) the way in which the above figure of 153 for meat and bacon could have been obtained in the first place and (ii) the way in which the food index you have obtained could then be combined with other indices to obtain the general index of retail prices.

34. The monthly accounts, to the nearest £, of 100 customers of a local store for February 1971 were as follows:

20	17	24	12	24	10	29	29	13	22
17	29	5	17	17	17	24	8	24	17
25	12	17	10	17	13	37	17	17	26
26	31	20	21	7	34	10	43	26	17
17	22	11	17	13	18	17	38	17	48
30	27	22	42	17	12	20	19	11	19
22	17	18	12	22	20	9	33	30	28
18	11	13	19	11	23	25	31	23	18
24	12	17	27	25	19	27	17	17	38
29	21	25	32	13	21	23	28	28	21

By arranging in groups $4\frac{1}{2}$–$9\frac{1}{2}$, $9\frac{1}{2}$–$14\frac{1}{2}$, ..., $44\frac{1}{2}$–$49\frac{1}{2}$, form a frequency distribution.

(i) Is the distribution Normal, bi-modal, skewed positively or skewed negatively?

(ii) What is its modal group?

(iii) What is its mode?

(iv) What is its median?

Diagrams are NOT required but a brief reason should be stated with each answer.

35. The following table, extracted from the *Monthly Digest of Statistics*, December 1970, shows the quarterly involuntary absence percentages of mine workers for the National Coal Board:

	Quarter			
Year	1	2	3	4
1967	—	—	—	12·0
1968	15·1	12·1	12·6	13·6
1969	15·3	12·4	12·5	13·5
1970	16·1	14·3	14·5	—

Calculate the four-quarterly moving averages and exhibit them, together with the original data, in a histogram.

Comment on the seasonal fluctuations and the trend as shown by your diagram.

36. (*a*) Describe THREE methods of measuring dispersion other than the standard deviation and discuss their relative merits.

(*b*) List the main points to be observed in designing a questionnaire for use in a statistical survey.

37. In the assembly of a machine the various components are arranged along the assembly line in a certain order. The following frequency distribution is a summary of the times taken for the assembly of 100 such machines.

Time, in minutes, of assembly (nearest minute)	Number of machines
38–40	10
41–43	24
44–46	38
47–49	23
50–52	4
53–55	1

Calculate the mean and the standard deviation of the times.

In the interests of efficiency of output it is decided to try arranging the various components in a new order and the times for the first 5 machines produced in this way are:

38, 50, 39, 42, 36 minutes.

Calculate the mean and the standard deviation of these 5 values and state what you infer about the new method compared with the old.

38. (*a*) Describe a table of random numbers with which you are familiar and explain how it can be used in sampling. Explain also how Market Research statistical enquiries are often carried out on a sampling basis without the use of random numbers or their equivalent.

(*b*) Explain how arithmetic probability paper differs from ordinary graph paper and how it can be used to demonstrate Normality and to estimate the mean and the standard deviation of a Normal distribution.

39. *Regional unemployment in the U.K.*

Region	Total unemployed 11 Jan. 1971	Rise since 7 Dec. 1970
South (E & W)	192,266	+22,820
Midlands (E & W) and East Anglia	113,075	+13,775
North	227,983	+18,675
Scotland	115,114	+11,598
Wales	42,269	+ 3,474
Ulster	40,853	+ 3,373
Total	731,560	+73,715

(*a*) Present the above unemployment figures for the six regions of the U.K. on 11 Jan. 1971 as a bar diagram and show the increases since 7 Dec. 1970 as suitably shaded sections of the bars.

(*b*) Suppose the total 731,560 for 11 Jan. 1971 is to be represented by a circle of radius 10 cm, the six regions being shown as sectors of the circle

(i) How many degrees should the sector representing 'North' contain?

(ii) What should the radius be for a circle representing the total unemployed on 7 Dec. 1970?

Show your calculations but DO NOT draw the circles.

40. The marks obtained by ten boys in an examination were

19, 25, 31, 40, 47, 48, 51, 55, 60, 74.

Calculate the standard deviation, correct to 2 decimal places. The marks were then adjusted so that the mean was 50 and the standard deviation 15. Calculate the highest and lowest marks on the new scale and obtain a formula for the new mark y in terms of the old mark x

Suggest any reason for making such an adjustment to examination marks. (A.E.B.)

41. The following table gives the 1948 and 1956 profits of six British companies:

	Net Profit (£ million)	
	1948	1956
British Petroleum	24·3	53·0
I.C.I.	11·2	27·1
Unilever (U.K. only)	10·3	22·1
British American Tobacco	8·9	21·5
Imperial Tobacco	8·0	12·3
Woolworth	4·5	11·1

In each case express the 1948 profit as a percentage of the 1956 profit.

Construct a bar diagram, using rectangles of width 20 mm, to represent the profits in 1956. By subdividing the *widths* of the rectangles, represent the 1948 profits as fractions of the 1956 profits. Comment on the progress made by the various companies. (J.M.B.)

42. *Receipts by Customs and Excise* (£ *million*)
Year ended 31 March 1956

Beer	259
Wines and spirits	150
Tobacco	669
Purchase tax	419
Entertainments and betting	68
Hydrocarbon oils	323
Other sources	122
Total	2010

Represent the above data by a circular diagram making the radius of the circle 50 mm.

Given that the total receipts for the previous year were £1870 million, calculate the radius of the circle which would effect a true comparison with the circle already drawn for 1956. (J.M.B.)

43. *Financial Times Index of Prices of Ordinary Shares*
(*Price in 1935 as 100*)

	1st Half-year	2nd Half-year
1951	—	135
1952	121	103
1953	113	131
1954	131	184
1955	175	224
1956	203	161
1957	207	159
1958	154	225
1959	212	338
1960	343	293
1961	363	340

By considering the above time-series in groups of ten, calculate the five-yearly moving averages and exhibit them on a graph together with a graph of the original series.

Comment on the trend shown by the moving averages, discussing whether or not investors have cause to be alarmed by the fluctuations during the last two years. (A.E.B.)

44. The following table shows, in £ thousand, the Passenger train receipts and the Coal and Coke freight train receipts of British Railways

for the 13 four-weekly periods of 1955 and the first 9 four-weekly periods of 1956:

Year and four-weekly period		Receipts for passengers (£ thousand)	Receipts for coal and coke (£ thousand)
1955	1	1683	2304
	2	1785	2350
	3	1900	2422
	4	2383	2202
	5	2001	2379
	6	1370	1400
	7	3142	2400
	8	3833	1829
	9	2865	2427
	10	2303	2604
	11	2048	2589
	12	1878	2604
	13	2285	2324
1956	1	1821	2474
	2	1811	2486
	3	2018	2570
	4	2407	2312
	5	2323	2611
	6	2489	2472
	7	3026	2439
	8	3779	1761
	9	2973	2265

On the same diagram, with the same axes and scales, draw two graphs to represent the two time-series.

Describe the fluctuations of each series, pointing out how they differ.

(J.M.B.)

45. The following frequency distribution was obtained from the results of comprehensive strength tests on 100 cubes of concrete:

Compressive strength (kg/mm²)	Number of cubes
2·250–2·499	1
2·500–2·749	7
2·750–2·999	26
3·000–3·249	38
3·250–3·499	22
3·500–3·749	6

Draw a cumulative frequency graph and use it to estimate the median and quartiles.

Calculate also, from the given table, the mean compressive strength, the standard deviation of the compressive strengths and the range of compressive strengths determined by the limits

mean $\pm$ two-thirds of the standard deviation.

(A.E.B.)

46. (*a*) The following frequency table was obtained in a test of 'distances to take-off' of a hundred aircraft of the same type:

Distance to take-off in metres	No. of aircraft
320–330	2
330–340	4
340–350	9
350–360	14
360–370	19
370–380	19
380–390	15
390–400	10
400–410	5
410–420	2
420–430	1

Construct a histogram to illustrate the data and calculate the mean distance to take-off.

(*b*) Find the median, the mode, the arithmetic mean and the geometric mean of the numbers

$$2,\quad 3,\quad 3,\quad 4,\quad 5,\quad 8,\quad 10.$$ (J.M.B.)

47. (*a*) If a certain population is represented by the area of a circle of radius R mm, find a formula for the radius r of a second circle which would represent a second population of size x % of the first.

(*b*) If a certain population is represented by the area of a circle, find a formula for the angle, in degrees, of a sector of the circle which represents y % of the population.

(*c*) Define the geometric mean of two different positive numbers a and b and prove that it is less than their arithmetic mean (J.M.B.)

48. *I.T.V. profits*

	1958 (£ million)	1959 estimate (£ million)
Associated Television (ATV)	3·689	5·150
Associated Rediffusion	5·108	7·000
Granada	1·160	
Associated British (ABC)	2·823	4·600

The above table shows the 1958 profits of four I.T.V. groups and estimates of the 1959 profits for three of the groups. If the proportional increase in profit of the Granada group is to equal the proportional increase in the total of the other three groups, obtain the missing 1959 estimate.

Taking the area of a square of side 20 mm to represent a profit of £1,000,000, draw on graph paper four squares to represent the 1958 profits of the four groups and label them appropriately. (J.M.B.)

49. *Consumption of cigarettes in the U.K.*

Year	Consumption of cigarettes in kg per head: Men	Women
1891	0·045	negligible
1901	0·45	negligible
1911	1·17	negligible
1921	2·16	negligible
1922	2·03	0·045
1923	2·11	0·045
1924	2·07	0·09
1926	2·25	0·09
1931	2·74	0·23
1936	3·20	0·36
1941	3·82	0·77
1946	4·30	1·13
1947	3·65	1·04
1948	3·40	0·90
1949	3·20	1·00
1950	3·20	1·04
1951	3·40	1·08
1956	3·60	1·40
1957	3·69	1·50
1958	3·74	1·50

On one diagram draw *two* graphs to illustrate the above time series.

Use your graphs to write an account of the variations in cigarette smoking by men and women. (J.M.B.)

50. *Number of students*

	1938–39	1949–50	1959–60	1969–70
Special Honours Chemistry	16	90	152	250
Other Pure Science	67	128	160	400
Medical and Dental	45	105	99	150
Engineering and Metallurgy	50	71	174	80
Total	178	394	585	880

The above table compares the numbers of students in a university Honours School of Chemistry with those in the other Pure Science Schools and in the faculties of Medicine and Engineering. Express each of the numbers as a percentage of the total for its year and present them in a self-explanatory table.

Construct a bar diagram to display simultaneously the yearly totals and the composition of each total.

Describe the calculations that would be necessary if, in an alternative diagram, the totals and their compositions were represented by circles. (You are not required to make the actual calculations or draw the circular diagrams.) (J.M.B.)

51. Taking 20 mm to represent 10 units on both the x and y scales, draw on one sheet of graph paper four separate diagrams to represent the data of tables (i), (ii), (iii) and (iv) given below.

(i)	x	5	9	12	16	19	2	5	7	12	14	18	3	7	11	14	17
	y	6	10	13	16	20	1	4	9	12	15	19	3	7	9	14	18
(ii)	x	5	9	13	16	18	9	12	14	2	6	7	15	18	5	7	3
	y	6	10	17	13	19	13	12	14	1	10	5	11	16	3	7	3
(iii)	x	4	11	13	16	17	4	15	12	19	5	15	16	19	2	4	4
	y	5	12	11	9	16	2	4	15	15	16	7	7	13	6	11	13
(iv)	x	3	6	9	11	13	15	1	13	15	18	17	8	2	6	8	4
	y	18	14	9	10	9	4	20	7	6	1	3	12	19	16	13	16

Comment, in each case, upon the kind of regression illustrated.

(J.M.B.)

52. (*a*) With the aid of a simple example in each case, show how you would determine for a set of observations (i) the arithmetic mean, (ii) the geometric mean, (iii) the median, (iv) the mode.

(*b*) The arithmetic mean of a set of n measurements is a. It is discovered that 2 of the measurements, x and y, are unreliable and must be discarded. Find an expression for the arithmetic mean of the remaining measurements. (J.M.B.)

53. (*a*) Four chain-base index numbers a, b, c and d are such that b is 95% of a, c is 85% of b and d is 110% of c. Express c and d as percentages of the fixed base a.

(*b*) The mean and the standard deviation of one random sample of 10 observations are 50 and 8 respectively. When a second random sample of 10 observations is combined with the first sample, the mean and the standard deviation of the combined sample of all 20 observations are 52 and 12 respectively.

Calculate the mean and the standard deviation of the second sample of 10 observations. (A.E.B.)

54. A certain type of car tyre was used on the front wheels of 250 cars of the same design and the number of kilometres covered by the tyres before replacement was recorded. The following table shows the results of the investigation:

Kilometres	No. of tyres
15,000–17,500	2
17,500–20,000	4
20,000–22,500	40
22,500–25,000	130
25,000–27,500	190
27,500–30,000	110
30,000–32,500	20
32,500–35,000	4
Total	500

Draw a cumulative frequency graph and use it to estimate the median and quartiles.

Calculate also, from the given table, the mean and the standard deviation of the kilometres.

Explain what inference you draw from (i) the quartiles, (ii) the standard deviation. (A.E.B.)

55. Sixty boys and forty girls took an examination. The means and the standard deviations of marks obtained by the two groups were as follows:

	Number	Mean	Standard deviation
Boys	60	48	12
Girls	40	53	10

Calculate the mean and the standard deviation of the combined group of one hundred candidates.

One of the candidates was awarded 100 marks and another 0 marks. If it were decided to ignore these two marks, calculate the mean and the standard deviation of the marks of the remaining 98 candidates.

56. The following table gives the deaths in the various age groups in a certain town:

Age group	0–5	5–15	15–25	25–55	over 55
Population of the town	p_1	p_2	p_3	p_4	p_5
Population of the whole country	w_1	w_2	w_3	w_4	w_5
Deaths in the town	d_1	d_2	d_3	d_4	d_5

Obtain the crude and standardized death-rates for the town and explain why the standardized death-rate is helpful. (A.E.B.)

57. The following numbers give the average weekly amount of electricity generated (in million kilowatt hours) by nuclear power stations for the first nine months of 1958.

5·6 5·9 6·8 11·4 9·6 10·7 3·9 2·5 3·0

Calculate the mean and the standard deviation of this set of numbers. (London.)

58. A pie chart is drawn to represent the monthly car sales on hire purchase for the period January to June, 1970. The angles of the six sectors of the chart are respectively 50°, 54°, 73°, 70°, 59°, 54°. If the total number of cars sold on hire purchase during the given period was 511,791, calculate the actual sales, correct to three significant figures, for each of the six months.

Calculate also the greatest deviation of monthly sales from the mean during this period. (London.)

59. The total index of industrial production is calculated as a weighted mean. From the following table calculate the rise in this index between 1955 and 1958 and express the rise as a percentage of the index in 1955.

		Index	
		1955	1958
	Weight	(Average 1954 =	100)
Total manufacturing	760	106	107
Mining and quarrying	72	99	94
Construction	120	100	105
Gas, electricity and water	48	105	117

(London.)

60. The following table gives the value (in million pounds) of exports and imports for each of the nine months during the period January to September, 1958. The exports are given as a deviation from £270 million and the imports as a deviation from £300 million.

	J	F	M	A	M	J	J	A	S
Imports	+24	−13	+17	+7	+5	−6	+34	+3	+9
Exports	+5	−8	+5	−8	+14	−49	+13	−15	−33

Calculate the coefficient of rank correlation between the values of exports and imports during this period.

Calculate also the total value of imports for the given period. (London.)

61. Calculate as a weighted mean, correct to the nearest integer, the total index of industrial production in January 1960 from the following table.

	Weight	Index (Average 1954 = 100)
Total manufacturing	760	121
Mining and quarrying	72	96
Construction	120	115
Gas, electricity and water	48	154

From 1960 to 1961 a rise of 5·6% in production in the 'Construction' industries produced a rise of 0·7% in the total index.

What rise in the total index would a rise of 3·2% in production in the 'gas, electricity and water' industries produce during the same period? (London.)

62. The following table gives the marks of 121 candidates in an examination. Construct a frequency table for these marks, using ten class intervals 0–9, 10–19, 20–29, ... 90–100. Draw a histogram of the distribution and comment on its nature.

34	64	36	51	35	48	23	30	25	31	58
44	41	55	39	47	56	62	63	50	66	42
21	50	61	33	38	66	62	55	49	39	48
46	33	46	45	58	51	64	56	48	45	41
42	41	34	48	48	38	47	21	45	24	20
55	22	25	30	56	13	19	18	14	24	45
28	17	17	21	17	28	24	32	36	45	28
61	40	37	44	71	43	38	36	31	43	56
19	29	34	26	45	50	34	50	47	41	36
43	51	81	63	65	50	59	75	50	62	69
76	51	60	83	64	40	61	82	56	64	59

(London.)

63. If r is the coefficient of correlation, explain carefully the significance of

$$\text{(i) } r = -0{\cdot}5, \quad \text{(ii) } r = +0{\cdot}9.$$

The following table gives the weekly average of coal production and the weekly average of gas available at gas works for nine months.

	Coal produced (Gg as unit, 3500 units as origin)	Gas produced (10^5 GJ as unit, 50 units as origin)
Apr.	+165	+ 4·8
May	+488	+ 0·6
June	+128	− 4·9
July	−522	− 6·4
Aug.	−549	− 7·4
Sept.	+202	− 1·1
Oct.	+380	+ 5·6
Nov.	+433	+11·8
Dec.	+157	+15·7

Calculate the coefficient of rank correlation between these items and comment on the result. (London.)

64. What factors must be taken into account when interpreting a coefficient of correlation and assessing its reliability?

The following table gives the monthly average value of imports and exports (in £ million) for the nine months April–December, 1960. The numbers are given as a deviation from £300 million.

	Import	Export
Apr.	+ 73·6	+ 8·9
May	+ 89·8	+18·8
June	+ 84·2	−23·8
July	+ 81·5	−23·6
Aug.	+ 81·5	−19·0
Sept.	+ 63·4	−38·8
Oct.	+ 78·7	−41·1
Nov.	+119·8	+44·0
Dec.	+ 83·3	+ 6·0

Calculate the coefficient of rank correlation between these items and comment on the result. (London.)

65. The following table gives the index numbers of wholesale prices (X) and of retail prices (Y) during the first eight months of 1961.

(1954 = 100)

	J	F	M	A	M	J	J	A
X	114·4	114·4	115·0	114·9	115·3	115·4	115·7	116·3
Y	112·3	112·3	112·7	113·3	113·6	114·6	114·6	115·7

(i) Explain the meaning of '1954 = 100'.

(ii) Plot a scatter diagram and draw a suitable straight line to represent the data.

(iii) Find the average change in Y corresponding to unit positive change in X. (London.)

66. Sales of footwear by a store in 52 consecutive weeks were:

37	60	67	63	69	54	68
60	62	83	66	70	68	61
74	94	87	66	69	66	98
62	78	90	47	70	68	98
40	73	93	51	70	71	
56	56	58	57	47	76	
59	64	46	53	54	67	
80	79	77	77	49	73	

Reduce these results to a frequency distribution with intervals centred at $40\frac{1}{2}$, $50\frac{1}{2}$, $60\frac{1}{2}$, etc., articles. Represent the results as a histogram drawn on a scale of 20 mm to each interval.

On a separate diagram plot the frequencies up to the ends of each interval and sketch a cumulative frequency curve. From this curve read off the median and quartiles of the distribution. (O. & C.)

67. Define the arithmetic mean $\bar{x}$ and the variance σ^2 of a set of n numbers $x_1, x_2, \ldots, x_n$. Prove that, if a is an arbitrary number and

$$\mu_2(a) = \frac{1}{n}\sum_{r=1}^{n}(x_r - a)^2,$$

then
$$\sigma^2 = \mu^2(a) - (\bar{x} - a)^2.$$

Ten rounds are fired from a gun at the same elevation and their ranges in metres, to the nearest ten metres are

7250	7280	7210	7290	7320
7190	7280	7250	7210	7270

Calculate the mean range, the standard deviation and the mean deviation. (O. & C.)

68. The breaking stresses in kg/mm^2 of 100 specimens taken from a large consignment of steel bars are given in the following grouped frequency distribution:

Centre of interval	39·6	39·8	40·0	40·2	40·4	40·6
Frequency	6	14	30	28	15	7

Calculate the mean and standard deviation.

Use the Normal distribution with this mean and standard deviation to estimate the proportion of bars that will break at a stress less than 39·75 kg/mm^2. (O. & C.)

69. (i) A pack of 52 cards contains 4 Aces and 4 Kings. Three cards are taken at random from the pack. Find the chance that they should be

(*a*) 3 Aces,

(*b*) 2 Aces and 1 King.

(ii) The chance of any one engine of a four-engined aeroplane failing on a long journey is 5%. If only one engine fails the chance of the aeroplane completing the journey is 80%; if two engines on opposite wings fail its chance of completing the journey is 50%. It cannot fly with two engines out of action on the same wing. Find the chance that the aeroplane will complete the journey.

(The probabilities given in this question bear no relation to those that may exist on regular air services.) (O. & C.)

70. Write down the Binomial distribution for the number of successes in n independent trials when the chance of success in a single trial is p. Prove that the mean of the distribution is np.

(i) In a trial 8 coins are tossed together. In 100 such trials how many times should one expect to obtain 3 'heads' and 5 'tails'?

(ii) If 8% of articles in a large consignment are defective, what is the chance that a sample of 30 articles will contain fewer than 3 defectives? (O. & C.)

71. Explain briefly what is meant by the standard error of the mean.

In a chemical plant the acid content of the effluent from the factory is measured frequently. From 400 measurements the acid content in grams per 100 litres of effluent is recorded in the following frequency distribution:

Acid content	12	13	14	15	16	Total
Frequency	5	52	235	74	34	400

(i) Find the mean acid content and the standard error of the mean.

(ii) Assuming a Normal distribution of the acid content, give 95% confidence limits for the mean acid content of the effluent.

(iii) Is the result consistent with a mean acid content of 14·13 grams per 100 litres obtained from tests over several years? (O. & C.)

72. The following table gives the number of road vehicles x (unit 1,000,000) in use in September of each year and the number of road casualties y (unit 100,000) in each year:

Year	Vehicles x	Casualties y
1952	4·9	2·1
1953	5·3	2·3
1954	5·8	2·4
1955	6·4	2·7
1956	6·9	2·7
1957	7·5	2·7
1958	7·9	3·0
1959	8·6	3·3

Find

(i) the covariance of x and y,

(ii) the line of regression of y on x,

(iii) the correlation coefficient between x and y. (O. & C.)

73. The gains in weight (in kg) of 100 pigs during a certain period had the following distribution:

Weight	No.	Weight	No.
0– 2	1	18–20	24
3– 5	1	21–23	13
6– 8	4	24–26	11
9–11	10	27–29	5
12–14	12	30–32	1
15–17	17	33–35	1

Calculate the mean and variance of this distribution.

Construct a histogram for this data, and mark on it the mean previously calculated. (J.M.B.)

74. (*a*) If all the sevens are removed from a normal pack of 52 cards, find the probability that a hand of six cards drawn from those remaining forms a sequence (i.e. with consecutive values, no account being taken of suit; an ace has the value one).

(*b*) If A and B each throw a pair of unbiased dice, show that the probability that they obtain the same pair of numbers is approximately 0·05. Find also the probability that they obtain the same total on the two dice. (J.M.B.)

75. The following table shows the lunch-time temperatures on the same day and the latitude north of the Equator, to the nearest half-degree, for a number of towns:

	Lunch-time temperature °C	Latitude north of the equator °
Amsterdam	4	$52\frac{1}{2}$
Barcelona	10	54
Belfast	2	$54\frac{1}{2}$
Berlin	3	52
Birmingham	10	$52\frac{1}{2}$
Bristol	10	$51\frac{1}{2}$
Brussels	6	51
Budapest	−5	$47\frac{1}{2}$
Edinburgh	1	56
Geneva	4	46
Helsinki	−4	60
Lisbon	6	39
London	9	$51\frac{1}{2}$
Madrid	4	$40\frac{1}{2}$
Moscow	−7	$55\frac{1}{2}$
New York	8	$40\frac{1}{2}$
Paris	7	$48\frac{1}{2}$
Rome	11	42
Vienna	4	48
Warsaw	−3	52

Calculate

(i) the coefficient of correlation and

(ii) the coefficient of correlation by ranks, for the two sets of figures.

(A.E.B.)

76. (*a*) In a certain subject the examination marks of the boys are Normally distributed about a mean 55 with standard deviation 11 and the marks of the girls are Normally distributed about a mean 58 with standard deviation 8. If one boy and one girl are chosen at random from the complete list of candidates determine

(i) the probability that the boy's mark is over 75,

(ii) the probability that the girl's mark is over 75,

(iii) the probability that the girl's mark is at least 20 more than the boy's,

(iv) the probability that the boy's mark is at least 20 more than the girl's.

(*b*) In another subject, the mean mark of a sample of 100 boys is 56 and that of 80 girls is 59. Assuming population standard deviations of 10 and 9 respectively determine whether the difference between these means is significant at the 5% level. (Mathematical Association.)

77. (i) At one stage in the assembly of a machine, a cylindrical rod with circular cross-section of diameter d has to fit into a circular socket of diameter D. Measurement of a large number of these rods and sockets indicates that both d and D are Normally distributed about respective means 5·01 cm and 5·11 cm with respective standard deviations 0·03 cm and 0·04 cm. If components are selected at random for assembly, show that 2·28 % of the rods are likely to be too big for the socket for which they are chosen.

(ii) In the production of machines described in (i) above, assuming that 2·28 % of the rods are too big for the socket for which they are chosen, calculate the probability that, in 100 random fittings, less than three rods will be too big. (Mathematical Association.)

78. The table below expresses in statistical terms a summary of a statement made in casual conversation by a man who travels regularly from the Victoria Air Terminal, London, to an office near the Empire State Building, New York.

	Mean time (hours)	Standard deviation (hours)
Bus journey from Victoria to Heathrow	0·5	0·1
Wait at Heathrow airport before take-off	1·5	0·3
Flight from Heathrow to the vicinity of Kennedy airport	7·0	0·5
Time waiting in the air for permission to land together with landing time	0·5	0·2
Wait at Kennedy airport	0·5	0·2
Bus journey from Kennedy to Empire State Building	0·5	0·2

Assuming that all the times are Normally and independently distributed show that the chance of travelling from Victoria to the Empire State Building in under 10 hours is less than a quarter.

(Mathematical Association.)

79. (*a*) When planning a ten-day holiday tour a motorist estimated the distance, x km, he would go each day by measuring the daily journey as a straight line on his map and taking x to the nearest 50. During the tour he recorded the actual daily distance y km. The

values x and y are tabulated below together with the ratios $z = y/x$. Calculate the mean value of z and the standard deviation.

Day	Estimated distance x km (nearest 50 km)	Actual distance y km (nearest integer)	Ratio $z = y/x$
1	250	305	1·22
2	150	243	1·62
3	100	162	1·62
4	250	290	1·16
5	100	149	1·49
6	250	295	1·18
7	300	387	1·29
8	400	448	1·12
9	250	325	1·30
10	250	375	1·50

For a *Normal* distribution of ratios $z = y/x$ with the same mean and standard deviation as the above sample find the probability of the actual distance being between 130 and 140 km when its estimate is 100 km.

(*b*) In order to test the hypothesis that ratios such as z are Normally distributed a teacher asked a hundred pupils to measure the distances, in metres, of their homes from school as straight lines on a map and then to measure the actual distances along the road. Describe the steps to be taken with the hundred values of z thus obtained to verify in some way that z is probably a Normal variable. Discuss whether or not a Normal distribution of z in the school experiment would justify the assumption of Normality in the case of the motorist. (Mathematical Association.)

80. The height, x mm, and the weight, y kg, of each boy in a school were recorded and a scatter diagram was then plotted. State, with reasons, which of the two possible regression lines is likely to prove the more useful.

The equation of the line of regression of y on x is found to be

$$y = 0{\cdot}08x - 80.$$

If the heights are distributed about a mean of 1600 mm with standard deviation 120 mm and the standard deviation of the weights is 12 kg find

(i) the mean of the weights,

(ii) the equation of the line of regression of x on y,

(iii) the coefficient of correlation. (Mathematical Association.)

81. The following table gives the quarterly index of industrial production over a four year period.

Year	1st quarter	2nd quarter	3rd quarter	4th quarter
First	115	113	106	116
Second	116	116	107	116
Third	116	117	109	118
Fourth	117	119	114	127

Calculate the four quarterly moving averages for the above data. Plot the original data and the moving averages on the same graph sheet using the same axes. Comment on any significant features.

(W.A.E.C.)

Glossary

Definitions of terms used in the text together with brief introductions to more advanced ideas.

Abscissa. The so-called 'horizontal' scale of a graph which is always used for the x-axis.

Absolute measure. A measure in terms of units.

Age-specific death-rate. The death-rate for a specified age-group.

Arithmetic mean. See 'mean'.

Average. A measure of central tendency of a group which enables us to assess the position in which the group stands with respect to other groups. A measure of location.

Bernoulli's theorem. The theorem of which the Binomial distribution is a corollary. If the probability of an event occurring at a single trial is p, the probability of exactly r occurrences of the event in n independent trials is

$$\binom{n}{r} p^r(1-p)^{n-r}.$$

Bimodal. A frequency distribution with two modes.

Binomial distribution. Suppose that the probability of an event occurring at a single trial is p and the probability of it not occurring is q then $p+q = 1$ and the probabilities

$$P(0), \quad P(1), \quad P(2), \quad \ldots, \quad P(r), \quad \ldots, \quad P(n)$$

of $0, 1, 2, \ldots, r, \ldots, n$ occurrences of the event in n independent trials are given by the terms of the *binomial expansion*

$$(q+p)^n = q^n + \binom{n}{1} q^{n-1}p + \binom{n}{2} q^{n-2}p^2 + \ldots + \binom{n}{r} q^{n-r}p^r + \ldots + p^n.$$

The mean value of $r, \mu = np$ and the variance of $r, \sigma^2 = npq$. As n increases the Binomial distribution tends to the normal. The minimum size of n for which the Normal distribution is a close approximation to the Binomial distribution depends on the value of p. The Normal approximation to the Binomial is quite good provided np and $n(1-p)$ are both greater than 5.

Birth-rate. The number of live births in any year per thousand of the estimated population at the middle of the year.

Central limit theorem. If random samples of size n and mean $\bar{x}$ are drawn from a population with mean μ and variance σ^2 the distribution of $(\bar{x}-\mu)/(\sigma/\sqrt{n})$ approaches *Normal* (0, 1) as n tends to infinity.

(Note that the approximation is quite good even when n is relatively small. For some parent populations $n \geqslant 15$ is necessary but for others $n \geqslant 5$ may be satisfactory.)

Central tendency. See 'average'.

Coefficient of correlation.

$$r_{xy} = \frac{\text{Covariance}}{\text{Product of Standard deviations}} = \frac{S_{xy}}{S_x . S_y}.$$

To test the significance of r_{xy} we use the t-distribution given in the *Cambridge Elementary Statistical tables*. r_{xy} differs significantly from zero (5% level) if

$$t = r_{xy}\sqrt{\left(\frac{n-2}{1-r_{xy}^2}\right)} \quad \text{with } \nu = n-2$$

is greater than the $P(=5\%)$ value of t.

For a given value of n, $r_{xy} = t/\sqrt{(t^2+n-2)}$ with $\nu = n-2$ gives the minimum values of r_{xy} in table 9 C.

Coefficient of rank correlation.

$$\text{Spearman's } \rho = 1 - \frac{6\Sigma D^2}{n(n^2-1)},$$

where D is the *rank difference* and n the *number of pairs*.

$$\text{Kendall's } \tau = \frac{S}{\frac{1}{2}n(n-1)},$$

where S is the *total score* for the various possible pairs and n the *number of pairs*.

Coefficient of regression. The gradient of the regression line.

The coefficient of regression of y on x is S_{xy}/S_x^2.

The coefficient of regression of x on y is S_{xy}/S_y^2.

Coefficient of variation.

$$\text{Coefficient of variation} = \frac{\text{Standard deviation}}{\text{Mean}}$$

$$= \frac{100 \times \text{Standard deviation}}{\text{Mean}}\%.$$

Confidence limits of the mean (large samples).

The 95% confidence limits are $m \pm \dfrac{1{\cdot}96\, s}{\sqrt{n}}$.

The 99·8% confidence limits are $m \pm \dfrac{3{\cdot}09\, s}{\sqrt{n}}$.

Continuous variation. A distribution in which values of the frequency may exist at every point of the abscissa scale. If may be represented graphically as a continuous curve, i.e. a frequency curve.

Corrected death-rate. A weighted average of age-specific death-rates in which the weights bear a definite relation to the age distribution of the population of the whole country.

Covariance. Covariance $= \dfrac{\Sigma d_x\, d_y}{n}$ where n is the number of (x, y) pairs and $\Sigma d_x\, d_y$ is the sum of the products of deviations from the mean.

Crude death-rate. The number of deaths in any year per thousand of the population alive at the middle of the year.

Discrete variation. A distribution in which values of the frequency only exist at certain separate points of the abscissa scale. It may be represented by a frequency polygon, but should not be represented by a continuous curve.

Geometric mean. The geometric mean of the n values $x_1, x_2, x_3, \ldots, x_n$ is

$$\sqrt[n]{(x_1 . x_2 . x_3 \ldots x_n)}.$$

Harmonic mean. The harmonic mean of the n values $x_1, x_2, x_3, \ldots, x_n$ is

$$\frac{1}{\dfrac{1}{n}\left\{\dfrac{1}{x_1}+\dfrac{1}{x_2}+\dfrac{1}{x_3}+\ldots+\dfrac{1}{x_n}\right\}}.$$

Infant mortality rate. The number of deaths of infants under one year of age per thousand live births.

Least squares line of best fit. If $y = a+bx$ is the equation, the values of a and b are given by the *normal* equations

$$\Sigma y = na+b\Sigma x,$$

$$\Sigma xy = a\Sigma x+b\Sigma x^2,$$

where n is the number of (x, y) pairs.

An alternative form of the equation is

$$(y-\bar{y}) = \frac{S_{xy}}{S_x^2}(x-\bar{x}),$$

or $$(y-\bar{y}) = \frac{r_{xy}\, s_y}{s_x}(x-\bar{x}).$$

Predictions of y based on the above equation can only be regarded as expectations. If α and β are the respective population values corresponding to a and b, the two-tail percentage points of the t-distribution, with $\nu = (n-2)$ degrees of freedom, can be used to obtain interval estimates of β, α and y. First calculate s_e, the standard error of estimate either by

$$s_e^2 = \frac{\sum_{r=1}^{n} (y_r - a - bx_r)^2}{n-2},$$

or by
$$s_e^2 = \frac{\Sigma y^2 - a\Sigma y - b\Sigma xy}{n-2}.$$

Then

(i) $t = \dfrac{|b-\beta|}{s_e}\sqrt{\left(\dfrac{n(\Sigma x^2)-(\Sigma x)^2}{n}\right)}$ gives the significance of $|b-\beta|$.

Note that b is said to be significant if it differs significantly from zero.

(ii) $b \pm \dfrac{ts_e}{\sqrt{\left(\dfrac{n(\Sigma x^2)-(\Sigma x)^2}{n}\right)}}$ gives the confidence limits for β.

(iii) $t = \dfrac{|a-\alpha|}{s_e\sqrt{\left(\dfrac{1}{n}+\dfrac{n\bar{x}^2}{n(\Sigma x^2)-(\Sigma x)^2}\right)}}$ gives the significance of $|a-\alpha|$.

Note that if a does not differ significantly from zero the line passes through the origin.

(iv) $a \pm ts_e\sqrt{\left(\dfrac{1}{n}+\dfrac{n\bar{x}^2}{n(\Sigma x^2)-(\Sigma x)^2}\right)}$ gives confidence limits for α.

(v) $(a+bx_0) \pm ts_e\sqrt{\left(\dfrac{1}{n}+\dfrac{n(x_0-\bar{x})^2}{n(\Sigma x^2)-(\Sigma x)^2}\right)}$ gives confidence limits for the *mean value of* y estimated from x_0.

(vi) $(a+bx_0) \pm ts_e\sqrt{\left(1+\dfrac{1}{n}+\dfrac{n(x_0-\bar{x})^2}{n(\Sigma x^2)-(\Sigma x)^2}\right)}$ gives confidence limits for an *individual value of* y estimated from x_0.

Location. See 'measure of location'.

Lower quartile. If n measurements are arranged in order of magnitude from the least to the greatest, the $\frac{1}{4}(n+1)$th is the lower quartile. The lower quartile divides the area of the histogram in the ratio 1: 3. The 25th percentile.

Mean. The arithmetic mean, or more simply the mean, of the n values $x_1, x_2, x_3, \ldots, x_n$ is

$$\frac{1}{n}\{x_1+x_2+x_3+\ldots+x_n\}.$$

Mean deviation. Mean deviation $= \frac{1}{n}\Sigma|d|$, where $n=$ the number of values and $\Sigma|d|=$ the sum of the moduli of their deviations from the mean, median or mode.

Measure of location. A single number which enables us to assess the position in which a group is located with respect to other groups. The mean, median and mode are examples of measures of location.

Median. If n measurements are arranged in order of magnitude the $\frac{1}{2}(n+1)$th is the median. The median bisects the area of the histogram. The 50th percentile.

Mode. In a frequency distribution the value which occurs most frequently is called the mode (or norm).

Modulus. The modulus of a number is its numerical value, no regard being paid to its + or − sign.

Negatively skewed distribution. A distribution in which more than half of the area of the histogram is to the left (negative) side of the mode. When the frequency curve has a long tail on the left and a short tail on the right it is said to be skewed negatively.

Normal frequency curve. The equation of the Normal frequency curve in its most general form is

$$y = \frac{1}{\sigma\sqrt{(2\pi)}}\,e^{-\frac{1}{2}(x-\mu)^2/\sigma^2}.$$

Here, μ is the mean and σ the standard deviation and the distribution is known as *Normal* (μ, σ^2). Table 7 A, page 71, gives values of the Normal (0, 1) distribution.

Null hypothesis. The null hypothesis is the *assumption* which is made when applying a significance test.

Ogive. The graph of a cumulative frequency distribution.

Ordinate. The so-called 'vertical' scale of a graph which is always used for the y-axis.

Parameters. Constants appearing in the specification of probability distributions, e.g. p (Binomial), a (Poisson), μ, σ^2 (Normal), λ (Exponential).

Percentiles. The values which divide the area of the histogram into a hundred equal parts.

Positively skewed distribution. A distribution in which more than half of the area of the histogram is to the right (positive) side of the mode. When the frequency curve has a long tail on the right and a short tail on the left it is said to be skewed positively.

Quartile deviation. Semi-interquartile range or quartile deviation

$$= \tfrac{1}{2}\,(\text{Upper quartile} - \text{Lower quartile}).$$

Quartiles. See 'Lower quartile' and 'Upper quartile'.

Relative measure. A ratio which is independent of units. A 'coefficient' is often, but not always, a relative measure.

Significance of a single mean (large samples)

If $\frac{|\mu_0 - m|}{S\sqrt{n}} > 1 \cdot 96$, m differs from μ_0 at the 5% level.

Significance of the difference between means (large samples)

If $\frac{|m_1 - m_2|}{\sqrt{\left(\frac{S_1^2}{n_1} + \frac{S_2^2}{n_2}\right)}} > 1 \cdot 96$, m_1 differs from m_2 at the 5% level.

Skewness. See 'positively skewed distribution' and 'negatively skewed distribution'.

Standard deviation. Standard deviation $= \sqrt{\left\{\frac{\Sigma d^2}{n}\right\}}$,

where $n =$ the number of values and $\Sigma d^2 =$ the sum of the squares of their deviations from the mean.

Standardized death-rate. See 'corrected death-rate'.

Upper quartile. If n measurements are arranged in order of magnitude from the least to the greatest, the $\frac{3}{4}(n+1)$th is the upper quartile. The upper quartile divides the area of the histogram in the ratio 3:1. The 75th percentile.

Answers to Exercises

§ 11. **1.** (i) 3·14, 2·72, 1·41, 1·73, 2·24, 2·45, 2·65, 3·16.
(ii) 3·1, 2·7, 1·4, 1·7, 2·2, 2·4, 2·6, 3·2.
(iii) 3, 3, 1, 2, 2, 2, 3, 3.

2. 15,000–24,999, 1500–2499.
It is wrong to have more significant figures in the total than in the separate items. 16,500–27,498.

3. 1, 4, 10, 18, 25, 22, 12, 6, 2. 177; 175–179.

4. 3, 7, 8, 10, 9, 13, 12, 12, 7, 3. 50–59; 63.

5. (i) 34–72; 29–83. (ii) 27; 26. (iii) 1; 6.

6. 39, 23, 14, 9, 6, 3, 2, 2, 1, 1.

13. 38, 31, 47, 42, 31, 37, 43, 47, 39, 45.

14. 1, 1, 2, 5, 8, 7, 6, 8, 1, 1.

§ 13. **1.** 74 kg. **2.** 1·06 m.

§ 24. **1.** 53, 70, $32\frac{1}{2}$. **2.** 25·3, 9·25, 2·72.
3. 1·08, 1·25, 0·90 m. **4.** 77·0, 85·1, 70·9 kg.
5. (i) 328; (ii) 315; (iii) 362, 246. **6.** (*a*) 15, 56; (*b*) 71, 31.
7. 70, 84, 46; 63, 73. **8.** (i) 45; (ii) 58; (iii) 59, 31.
9. £4·55, £5·25, £3·75. **10.** 33·2, 34·8, 31·0.
11. 908, 943, 868; 1714, 1761, 1667 mm.
12. 5, 11, 9, 7, 3, 1, 0; mode = 1; median = 2.

§ 26. **1.** 981·2. **2.** 42·4. **3.** 100·1.

§ 29. **2.** 59·2 kg.

§ 32. **3.** 77·8 kg. **4.** 1·076 m. **5.** 21 min. **6.** 302·7.
7. £4·49. **8.** 907 mm, 1714 mm.
9. Make divisions at 140·5, 145·5 etc., mode = 153·3; mean = 153·4.
Mean gives consideration to everyone; mode does not.

§ 41. **1.** $222\frac{2}{9}$ km/h. **2.** $10\frac{10}{11}$ km/litre.

§ 48. **1.** 0·25 mm, 0·09 mm. **2.** 0·63. **3.** 7·6. **4.** 100·2, 10·3.
5. 9·1, 4·8. **6.** 6·71, 9·93.

§ 50. **1.** 68·63. **2.** 12·45, 13·05. **3.** 22·0. **4.** £0·828. **5.** 2·36.

§ 55. **1.** 0·74. **2.** 10·25. **3.** 1·35. **4.** 10·9, 5·66.
5. 7·22, 11·97.

§ 60. **1.** 88·2. **2.** 15·1, 16·2. **3.** £1·09. **4.** 3·26.
5. 0·06, 0·04. **6.** 16·7%, 18%. **7.** 15·6%, 20%.

§ 71. **1.** $\frac{11}{24}$. **2.** (i) $\frac{23}{100}$; (ii) $\frac{379}{1800}$. **3.** 0·46.
4. $\frac{1}{120}$. **5.** (i) $\frac{1}{1296}$; (ii) $\frac{19}{1296}$; (iii) $\frac{523}{648}$.
6. (*a*) $\frac{7}{16}$; (*b*) $\frac{9}{13}$; (*c*) $\frac{9}{40}$.

§ 79. **1.** $\frac{1}{16}$, $\frac{4}{16}$, $\frac{6}{16}$, $\frac{4}{16}$, $\frac{1}{16}$; 2·5, 10, 15, 10, 2·5.

§ 81. **1.** 59·0, 32·8, 7·3, 0·9. **2.** 60·3, 30·9, 7·5, 1·3.

§ 84. **1.** (i) 0·15; (ii) 0·96. **2.** 4, 11, 14, 11, 6, 3, 1.
3. (i) 0·0055; (ii) 0·2903. **4.** (i) $\frac{864}{3125}$; (ii) $\frac{513}{625}$.
5. $\frac{280}{2187}$ = 0·128.

§ 89. **1.** 0·24, 0·048. **2.** $18\frac{3}{4}$%.

§ 99. **1.** (i) 0·0228; (ii) 0·1587. **2.** 2·28.
3. (i) 0·62%; (ii) 100·5 ± 0·135 mm.
4. 0·03, 0·79, 7·26, 26·38, 38·11, 21·95, 5·01, 0·47.
5. 1, 7, 26, 38, 22, 5, 1. **6.** (i) 6·68; (ii) 0·023.
7. (i) 5·48; (ii) 0·82.
8. 60 mm; 0·2 mm. (i) 6·68% over and under; (ii) 0·62% over and under.
9. 100 ± 8 ohms.

§ 110. **1.** French and German marks are directly correlated. German marks about 20 better than French.
2. (*a*) Botany and Zoology marks are directly correlated. (*b*) Marks are about equal in each subject.
3. 19·3 cm.
4. (*a*) Physics marks about 15 better than Mathematics. (*b*) Physics and Mathematics marks directly correlated.
5. 0·12. **6.** 1·7. **8.** *O* is farthest above the line.
9. 12·5; 11·4 along 10·8 across. **11.** 26, 33, 43, 52, 61, 71; 58 years.
12. 34, 34, 37·8, 41·6, 49·4, 55·3, 61·6, 62·8; 580 kg.
13. Yes; points are all close to $y = 3{\cdot}56x - 3{\cdot}25$.

§ 116. **1.** 0·98. **2.** 0·98. **3.** Unity. **4.** 0·97.
5. 0·32. **6.** 0·86. **7.** −0·93. **8.** 0·83.
9. 0·86. **10.** 0·82. **11.** 0·91. **12.** 0·34.
13. 0·98. **14.** 0·89. **15.** 0·92.
16. 10·12, 337·2; 2·46, 44·5, 102·3; 0·93.

§ 121. **1.** 0·96, 0·90. **2.** 0·98, 0·93. **3.** Unity, unity.
4. 0·90, 0·76. **5.** 0·26, 0·17. **6.** 0·89, 0·76.
7. −0·95, −0·83. **8.** 0·82, 0·65. **9.** 0·81, 0·65.
10. 0·80, 0·60. **11.** 0·91, 0·80. **12.** 0·37, 0·22.
13. 0·997, 0·98. **14.** 0·98, 0·89. **15.** 0·95, 0·84.
16. 0·78. **17.** 0·8, 0·64. **18.** 0·02, 0·07.
19. 0·62, 0·42. **20.** 0·67; correlation probable (table 10 C).

§ 125. **1.** $73\frac{3}{4}$, $74\frac{1}{2}$, $75\frac{1}{4}$, $75\frac{1}{2}$, $76\frac{1}{2}$, $77\frac{3}{4}$, $79\frac{1}{2}$, 81, $82\frac{1}{4}$.
2. 86·4, 86·525, 89·175, 90·425, 91·575, 92·525, 90·0, 88·125, 87·025.
3. 11·275, 11·075, 11·925, 12·425, 13·050, 13·175, 11·800, 10·625, 9·700, 9·575, 9·825, 10·375, 10·850.
4. $24\frac{1}{4}$, 25, 25, 28, $30\frac{3}{4}$, $33\frac{1}{2}$, $37\frac{1}{2}$, $39\frac{1}{4}$, 38.
5. $72\frac{3}{4}$, $74\frac{3}{4}$, $76\frac{1}{4}$, $76\frac{3}{4}$, $77\frac{1}{2}$, $79\frac{1}{2}$, $82\frac{1}{4}$, $84\frac{1}{2}$, $87\frac{1}{4}$.

6. $323\frac{1}{4}$, $314\frac{3}{4}$, $311\frac{1}{4}$, $316\frac{1}{2}$, $327\frac{1}{4}$, $349\frac{1}{2}$, $367\frac{1}{4}$, $376\frac{1}{2}$, $385\frac{3}{4}$, 394, $402\frac{1}{4}$, $409\frac{1}{2}$, $416\frac{1}{4}$, $418\frac{3}{4}$, 423, $428\frac{1}{4}$, $432\frac{1}{2}$, $438\frac{1}{4}$, $444\frac{3}{4}$, 451, $457\frac{1}{2}$.

§ 130. **1.** $20\frac{1}{12}$, $19\frac{11}{12}$, 20, $20\frac{1}{12}$, $19\frac{11}{12}$, $20\frac{1}{4}$, $20\frac{1}{3}$, $20\frac{5}{6}$, $21\frac{3}{4}$, $22\frac{1}{2}$, $23\frac{2}{3}$, $24\frac{5}{6}$, $26\frac{1}{12}$, 27, $28\frac{1}{12}$, $29\frac{1}{4}$.

2. (In hundreds) 175, $178\frac{7}{12}$, 181, 186, 190, $194\frac{7}{12}$, $199\frac{1}{6}$, $204\frac{3}{4}$, $212\frac{1}{3}$.

3. 17·5, 19·2, 21·2, 23, 25, 26·5, 28·2, 29·5, 30·5, 32, 33·2, 34·8, 35·8, 37·5, 39, 39, 39·3, 40, 40, 39·7, 39·5, 39·5, 39·5, 39·5, 39·5, 38·5, 37·5, 36·5, 35·2, 33·5, 32·8.

7. 81·9, 83·9, 85·4, 86·6, 87·5, 88·9, 89·7, 91·7, 93·1, 94·1, 95·2.

8. 73·8, 72·4, 71·0, 70·0, 67·4, 64·8, 61·0, 58·4, 57·6, 57·8 (beef), 92·4, 89·6, 87·6, 83·0, 78·4, 75·4, 72·0, 70·8, 73·8, 72·4 (mutton).

§ 138. **1.** 144. **2.** 114. **3.** 114.

4. (*a*) 109. (*b*) 90, 95, 94. **5.** $66\frac{2}{3}$. **6.** 117·1.

7. 13·0; 14·3. **8.** 19·16, 6·56, 16·09; 7·04.

9. 1970 price relatives are 150, 200, 120, 125, 140; weighted average = 152·5.

§ 150. 5 out of 5.

§ 154. **1.** 0·0173, 0·0867, 0·195. No; probability 10·4% of equally good result.

2. (i) 0·0463; (ii) 0·00168. Yes; 4·14%.

4. (i) 0·026, 0·130, 0·197; (ii) 0·027, 0·121, 0·199.

5. Not; but 62 would be. **6.** Improvement sig. at 5% level. Compatible.

7. No; 40% to 60%. **8.** 52; No. **9.** 0·2; No.

10. (i) Yes, 2·2%; (ii) 5.

§ 158. **1.** (i) $1·04 \pm 0·018$; (ii) $1·04 \pm 0·028$; 0·68.

2. (i) $7·45 \pm 0·178$; (ii) $7·45 \pm 0·281$; 0·49.

§ 161. **1.** Yes; 2·36. **2.** No; 1·77. **3.** No; 1·67.

4. 1216; 75. No; not sig. because $16/10·71 < 1·96$.

§ 166. **1.** 13·70, 0·5093. **2.** 3 min 49 sec; 2·015 sec; 0·0236.

3. 0·06, 0·0384; 79·2%.

§ 171. **1.** 2·24; sig. at 5% level but not at 1% level.

2. 2·12; difference is significant. Also the standard deviations prove that the thickness is nowhere less than 5 mm.

Chapter 15 Miscellaneous Exercises

2. £4·98, £8·15. (*a*) B £2·65, D £0·25. (*b*) A £1·75, C £1·15.

5. 20·3, 20·7, 20·3, 20·5, 20·3, 20·5, 20·4, 20·4, 20·6, 20·8; 21·3, 21·7, 22·0.

6. $42\frac{1}{2}$. **7.** 42·3, 7·8. **8.** (i) 42 mm; (ii) 11 mm.

9. X better than Y. **10.** 0·76, 0·60.

12. 2890×10^6 kg. Between 2885×10^6 and 2895×10^6 kg.

15. (i) 3·26; (ii) 5·6.

16. $230\frac{4}{9}$; 54, 41, 35.

17. 1730 mm, 1710 mm.

18. 1735 mm, 62·5 mm.

19. $111\frac{3}{4}$, $114\frac{1}{2}$, 117, 119, $121\frac{1}{4}$, 124, $125\frac{1}{4}$, $126\frac{3}{4}$, $130\frac{1}{4}$, $135\frac{1}{4}$.

20. 0·88, 0·73.

21. 30 mm.

22. (*a*) 72%, 86·4%. (*b*) $(n_1 x + n_2 y)/(n_1 + n_2)$.

24. 13·6, 24·0.

26. 62·1; (i) 65; (ii) 91.

27. 873, 796 seats.

30. 30·28 km/h; 24 km/h.

31. 125·99.

32. 132·46.

33. 143·3.

34. (i) Skewed positively; (ii) $14\frac{1}{2}$–$19\frac{1}{2}$; (iii) 17; (iv) 20·5.

35. 12·95, 13·35, 13·40, 13·475, 13·45, 13·425, 13·625, 14·1, 14·6.

37. 44·7 min, 3·16 min; 41·0 min, 4·9 min. New method quicker but more erratic. May become less variable when the workers have got used to it.

39. 112°; 9·48 cm.

40. 15·85, 77·4, 25·4; $y = 0{\cdot}946x + 7{\cdot}4$.

41. 45·8, 41·3, 46·6, 41·4, 65·0, 40·5.

42. 48·2 mm.

43. 152·0, 154·6, 163·2, 168·8, 172·9, 182·3, 190·4, 205·8, 222·6, 229·5, 245·5, 263·4.

45. 3·1; 2·9; 3·3. 3·1; 0·26. 2·92 to 3·28.

46. (*a*) 371·3; (*b*) 4, 3, 5, 4·34.

47. $R\sqrt{x}/10$; $3{\cdot}6y$.

48. £1·672 million.

50.

9·0	22·8	26·0	28·4
37·6	32·5	27·4	45·5
25·3	26·7	16·9	17·0
28·1	18·0	29·7	9·1

52. $(an - x - y)/(n - 2)$.

53. (*a*) 80·7, 88·8; (*b*) 54, 14·7.

54. 25,970; 24,060; 27,700. 25,910; 2680.

55. 50; 11·5. 50; 9·14 to 9·17.

56. $1000\{\Sigma d/\Sigma p\}$; $1000\{\Sigma(wd/p)\}/\Sigma w$.

57. 6·6; 3·12.

58. 71,100, 76,800, 104,000, 99,500, 83,900, 76,800; 18,650.

59. $106{\cdot}3 - 104{\cdot}7 = 1{\cdot}6$; 1·5.

60. $\rho = 0{\cdot}52$; $\tau = 0{\cdot}39$. Total value £2780 million. Correlation not probable (tables 10 C, 10 G).

61. 120; 0·7 (48/120) (3·2/5·6) = 0·16.

62. 0, 8, 16, 22, 30, 22, 17, 3, 3, 0. Normal.

63. $\rho = 0{\cdot}62$; $\tau = 0{\cdot}50$. Correlation not probable (tables 10 C, 10 G).

64. $+ve$ or $-ve$? Near 1 or 0? Is it spurious? Is it a representative sample? $\rho = 0{\cdot}61$; $\tau = 0{\cdot}47$. Correlation not probable (tables 10 C, 10 G).

65. (iii) $S_{xy}/S_x^2 = 0{\cdot}6709/0{\cdot}3644 = 1{\cdot}84$.

66. 2, 8, 12, 17, 7, 4, 2. 67·6; 57·7; 76·1.

67. 7255 m; 39 m. 33 m.

68. 40·106; 0·254. $A(-1{\cdot}40) = 8\%$.

69. (i) (*a*) $\frac{1}{5525}$; (*b*) $\frac{6}{5525}$.

(ii) $(0\cdot95)^4+4(0\cdot95)^3(0\cdot05)(0\cdot80)+4(0\cdot95)^2(0\cdot05)^2(0\cdot50) = 0\cdot833$.

70. (i) about 22; (ii) 0·57.

71. (i) 14·2; 0·0408. (ii) $14\cdot2 \pm 0\cdot08$. (iii) Yes.

72. (i) 0·4244. (ii) $y = 0\cdot29x+0\cdot74$. (iii) 0·97.

73. 17·83; 36·7.

74. (*a*) $\frac{4}{48}.\frac{4}{47}.\frac{4}{46}.\frac{4}{45}.\frac{4}{44}.\frac{4}{43}.(6.5.4.3.2.1.)\,2 = 0\cdot00067$.

(*b*) $\frac{1}{18} = 0\cdot05$ approx. $\{2(1^2+2^2+\ldots+5^2)+6^2\}/36^2 = 0\cdot113$.

75. (i) $S_x = 10\cdot2$; $S_y = 5\cdot58$; $S_{xy} = -22\cdot8$; $r_{xy} = -0\cdot385$.

(ii) $\rho = -0\cdot36$. (ii) $\tau = -0\cdot24$.

Probability of correlation less than 5% (Tables 9 C, 10 C, 10 G).

76. (*a*) (i) 0·0344; (ii) 0·0168; (iii) 0·1056; (iv) 0·0455.

(*b*) Significant at 5% level (2·14).

77. 0·60. **78.** 0·233.

79. (*a*) 1·35, 0·18; 22%.

(*b*) Plot cumulative distribution on probability paper.

No. differences over short journeys not the same as differences over long journeys.

80. (i) 48 kg; (ii) $x = 8y+1216$; (iii) 0·8.

81. 112·5, 112·75, 113·5, 113·75, 113·75, 113·75, 114·0, 114·5, 115·0, 115·25, 115·75, 117·0, 119·25.

Seasonal fluctuation; upward trend.

Index

	0	1	2	3	4	5	6	7	8	9	Differences 1	2	3	4	5	6	7	8
10	0000	0043	0086	0128	0170	0212	0253	0294	0334	0374	4	8	12	17	21	25	29	33
11	0414	0453	0492	0531	0569	0607	0645	0682	0719	0755	4	8	11	15	19	23	26	30
12	0792	0828	0864	0899	0934	0969	1004	1038	1072	1106	3	7	10	14	17	21	24	28
13	1139	1173	1206	1239	1271	1303	1335	1367	1399	1430	3	6	10	13	16	19	23	26
14	1461	1492	1523	1553	1584	1614	1644	1673	1703	1732	3	6	9	12	15	18	21	24
15	1761	1790	1818	1847	1875	1903	1931	1959	1987	2014	3	6	8	11	14	17	20	22
16	2041	2068	2095	2122	2148	2175	2201	2227	2253	2279	3	5	8	11	13	16	18	21
17	2304	2330	2355	2380	2405	2430	2455	2480	2504	2529	2	5	7	10	12	15	17	20
18	2553	2577	2601	2625	2648	2672	2695	2718	2742	2765	2	5	7	9	12	14	16	19
19	2788	2810	2833	2856	2878	2900	2923	2945	2967	2989	2	4	7	9	11	13	16	18
20	3010	3032	3054	3075	3096	3118	3139	3160	3181	3201	2	4	6	8	11	13	15	17
21	3222	3243	3263	3284	3304	3324	3345	3365	3385	3404	2	4	6	8	10	12	14	16
22	3424	3444	3464	3483	3502	3522	3541	3560	3579	3598	2	4	6	8	10	12	14	15
23	3617	3636	3655	3674	3692	3711	3729	3747	3766	3784	2	4	6	7	9	11	13	15
24	3802	3820	3838	3856	3874	3892	3909	3927	3945	3962	2	4	5	7	9	11	12	14
25	3979	3997	4014	4031	4048	4065	4082	4099	4116	4133	2	3	5	7	9	10	12	14
26	4150	4166	4183	4200	4216	4232	4249	4265	4281	4298	2	3	5	7	8	10	11	13
27	4314	4330	4346	4362	4378	4393	4409	4425	4440	4456	2	3	5	6	8	9	11	13
28	4472	4487	4502	4518	4533	4548	4564	4579	4594	4609	2	3	5	6	8	9	11	12
29	4624	4639	4654	4669	4683	4698	4713	4728	4742	4757	1	3	4	6	7	9	10	12
30	4771	4786	4800	4814	4829	4843	4857	4871	4886	4900	1	3	4	6	7	9	10	11
31	4914	4928	4942	4955	4969	4983	4997	5011	5024	5038	1	3	4	0	7	8	10	11
32	5051	5065	5079	5092	5105	5119	5132	5145	5159	5172	1	3	4	5	7	8	9	11
33	5185	5198	5211	5224	5237	5250	5263	5276	5289	5302	1	3	4	5	6	8	9	10
34	5315	5328	5340	5353	5366	5378	5391	5403	5416	5428	1	3	4	5	6	8	9	10
35	5441	5453	5465	5478	5490	5502	5514	5527	5539	5551	1	2	4	5	6	7	9	10
36	5563	5575	5587	5599	5611	5623	5635	5647	5658	5670	1	2	4	5	6	7	8	10
37	5682	5694	5705	5717	5729	5740	5752	5763	5775	5786	1	2	3	5	6	7	8	9
38	5798	5809	5821	5832	5843	5855	5866	5877	5888	5899	1	2	3	5	6	7	8	9
39	5911	5922	5933	5944	5955	5966	5977	5988	5999	6010	1	2	3	4	5	7	8	9
40	6021	6031	6042	6053	6064	6075	6085	6096	6107	6117	1	2	3	4	5	6	8	9
41	6128	6138	6149	6160	6170	6180	6191	6201	6212	6222	1	2	3	4	5	6	7	8
42	6232	6243	6253	6263	6274	6284	6294	6304	6314	6325	1	2	3	4	5	6	7	8
43	6335	6345	6355	6365	6375	6385	6395	6405	6415	6425	1	2	3	4	5	6	7	8
44	6435	6444	6454	6464	6474	6484	6493	6503	6513	6522	1	2	3	4	5	6	7	8
45	6532	6542	6551	6561	6571	6580	6590	6599	6609	6618	1	2	3	4	5	6	7	8
46	6628	6637	6646	6656	6665	6675	6684	6693	6702	6712	1	2	3	4	5	6	7	7
47	6721	6730	6739	6749	6758	6767	6776	6785	6794	6803	1	2	3	4	5	5	6	7
48	6812	6821	6830	6839	6848	6857	6866	6875	6884	6893	1	2	3	4	4	5	6	7
49	6902	6911	6920	6928	6937	6946	6955	6964	6972	6981	1	2	3	4	4	5	6	7
50	6990	6998	7007	7016	7024	7033	7042	7050	7059	7067	1	2	3	3	4	5	6	7
51	7076	7084	7093	7101	7110	7118	7126	7135	7143	7152	1	2	3	3	4	5	6	7
52	7160	7168	7177	7185	7193	7202	7210	7218	7226	7235	1	2	2	3	4	5	6	7
53	7243	7251	7259	7267	7275	7284	7292	7300	7308	7316	1	2	2	3	4	5	6	6
54	7324	7332	7340	7348	7356	7364	7372	7380	7388	7396	1	2	2	3	4	5	6	6
	0	**1**	**2**	**3**	**4**	**5**	**6**	**7**	**8**	**9**	**1**	**2**	**3**	**4**	**5**	**6**	**7**	**8**

0	1	2	3	4	5	6	7	8	9	Differences								
										1	2	3	4	5	6	7	8	9
7404	7412	7419	7427	7435	7443	7451	7459	7466	7474	1	2	2	3	4	5	5	6	7
7482	7490	7497	7505	7513	7520	7528	7536	7543	7551	1	2	2	3	4	5	5	6	7
7559	7566	7574	7582	7589	7597	7604	7612	7619	7627	1	2	2	3	4	5	5	6	7
7634	7642	7649	7657	7664	7672	7679	7686	7694	7701	1	1	2	3	4	4	5	6	7
7709	7716	7723	7731	7738	7745	7752	7760	7767	7774	1	1	2	3	4	4	5	6	7
7782	7789	7796	7803	7810	7818	7825	7832	7839	7846	1	1	2	3	4	4	5	6	6
7853	7860	7868	7875	7882	7889	7896	7903	7910	7917	1	1	2	3	4	4	5	6	6
7924	7931	7938	7945	7952	7959	7966	7973	7980	7987	1	1	2	3	3	4	5	6	6
7993	8000	8007	8014	8021	8028	8035	8041	8048	8055	1	1	2	3	3	4	5	5	6
8062	8069	8075	8082	8089	8096	8102	8109	8116	8122	1	1	2	3	3	4	5	5	6
8129	8136	8142	8149	8156	8162	8169	8176	8182	8189	1	1	2	3	3	4	5	5	6
8195	8202	8209	8215	8222	8228	8235	8241	8248	8254	1	1	2	3	3	4	5	5	6
8261	8267	8274	8280	8287	8293	8299	8306	8312	8319	1	1	2	3	3	4	5	5	6
8325	8331	8338	8344	8351	8357	8363	8370	8376	8382	1	1	2	3	3	4	4	5	6
8388	8395	8401	8407	8414	8420	8426	8432	8439	8445	1	1	2	2	3	4	4	5	6
8451	8457	8463	8470	8476	8482	8488	8494	8500	8506	1	1	2	2	3	4	4	5	6
8513	8519	8525	8531	8537	8543	8549	8555	8561	8567	1	1	2	2	3	4	4	5	5
8573	8579	8585	8591	8597	8603	8609	8615	8621	8627	1	1	2	2	3	4	4	5	5
8633	8639	8645	8651	8657	8663	8669	8675	8681	8686	1	1	2	2	3	4	4	5	5
8692	8698	8704	8710	8716	8722	8727	8733	8739	8745	1	1	2	2	3	4	4	5	5
8751	8756	8762	8768	8774	8779	8785	8791	8797	8802	1	1	2	2	3	3	4	5	5
8808	8814	8820	8825	8831	8837	8842	8848	8854	8859	1	1	2	2	3	3	4	5	5
8865	8871	8876	8882	8887	8893	8899	8904	8910	8915	1	1	2	2	3	3	4	4	5
8921	8927	8932	8938	8943	8949	8954	8960	8965	8971	1	1	2	2	3	3	4	4	5
8976	8982	8987	8993	8998	9004	9009	9015	9020	9025	1	1	2	2	3	3	4	4	5
9031	9036	9042	9047	9053	9058	9063	9069	9074	9079	1	1	2	2	3	3	4	4	5
9085	9090	9096	9101	9106	9112	9117	9122	9128	9133	1	1	2	2	3	3	4	4	5
9138	9143	9149	9154	9159	9165	9170	9175	9180	9186	1	1	2	2	3	3	4	4	5
9191	9196	9201	9206	9212	9217	9222	9227	9232	9238	1	1	2	2	3	3	4	4	5
9243	9248	9253	9258	9263	9269	9274	9279	9284	9289	1	1	2	2	3	3	4	4	5
9294	9299	9304	9309	9315	9320	9325	9330	9335	9340	1	1	2	2	3	3	4	4	5
9345	9350	9355	9360	9365	9370	9375	9380	9385	9390	1	1	2	2	3	3	4	4	5
9395	9400	9405	9410	9415	9420	9425	9430	9435	9440	0	1	1	2	2	3	3	4	4
9445	9450	9455	9460	9465	9469	9474	9479	9484	9489	0	1	1	2	2	3	3	4	4
9494	9499	9504	9509	9513	9518	9523	9528	9533	9538	0	1	1	2	2	3	3	4	4
9542	9547	9552	9557	9562	9566	9571	9576	9581	9586	0	1	1	2	2	3	3	4	4
9590	9595	9600	9605	9609	9614	9619	9624	9628	9633	0	1	1	2	2	3	3	4	4
9638	9643	9647	9652	9657	9661	9666	9671	9675	9680	0	1	1	2	2	3	3	4	4
9685	9689	9694	9699	9703	9708	9713	9717	9722	9727	0	1	1	2	2	3	3	4	4
9731	9736	9741	9745	9750	9754	9759	9763	9768	9773	0	1	1	2	2	3	3	4	4
9777	9782	9786	9791	9795	9800	9805	9809	9814	9818	0	1	1	2	2	3	3	4	4
9823	9827	9832	9836	9841	9845	9850	9854	9859	9863	0	1	1	2	2	3	3	4	4
9868	9872	9877	9881	9886	9890	9894	9899	9903	9908	0	1	1	2	2	3	3	4	4
9912	9917	9921	9926	9930	9934	9939	9943	9948	9952	0	1	1	2	2	3	3	4	4
9956	9961	9965	9969	9974	9978	9983	9987	9991	9996	0	1	1	2	2	3	3	3	4
0	1	2	3	4	5	6	7	8	9	1	2	3	4	5	6	7	8	9

	0	1	2	3	4	5	6	7	8	9	Differences 1	2	3	4	5	6	7	8
·00	1000	1002	1005	1007	1009	1012	1014	1016	1019	1021	0	0	1	1	1	1	2	2
·01	1023	1026	1028	1030	1033	1035	1038	1040	1042	1045	0	0	1	1	1	1	2	2
·02	1047	1050	1052	1054	1057	1059	1062	1064	1067	1069	0	0	1	1	1	1	2	2
·03	1072	1074	1076	1079	1081	1084	1086	1089	1091	1094	0	0	1	1	1	1	2	2
·04	1096	1099	1102	1104	1107	1109	1112	1114	1117	1119	0	1	1	1	1	2	2	2
·05	1122	1125	1127	1130	1132	1135	1138	1140	1143	1146	0	1	1	1	1	2	2	2
·06	1148	1151	1153	1156	1159	1161	1164	1167	1169	1172	0	1	1	1	1	2	2	2
·07	1175	1178	1180	1183	1186	1189	1191	1194	1197	1199	0	1	1	1	1	2	2	2
·03	1202	1205	1208	1211	1213	1216	1219	1222	1225	1227	0	1	1	1	1	2	2	2
·09	1230	1233	1236	1239	1242	1245	1247	1250	1253	1256	0	1	1	1	1	2	2	2
·10	1259	1262	1265	1268	1271	1274	1276	1279	1282	1285	0	1	1	1	1	2	2	2
·11	1288	1291	1294	1297	1300	1303	1306	1309	1312	1315	0	1	1	1	2	2	2	2
·12	1318	1321	1324	1327	1330	1334	1337	1340	1343	1346	0	1	1	1	2	2	2	2
·13	1349	1352	1355	1358	1361	1365	1368	1371	1374	1377	0	1	1	1	2	2	2	3
·14	1380	1384	1387	1390	1393	1396	1400	1403	1406	1409	0	1	1	1	2	2	2	3
·15	1413	1416	1419	1422	1426	1429	1432	1435	1439	1442	0	1	1	1	2	2	2	3
·16	1445	1449	1452	1455	1459	1462	1466	1469	1472	1476	0	1	1	1	2	2	2	3
·17	1479	1483	1486	1489	1493	1496	1500	1503	1507	1510	0	1	1	1	2	2	2	3
·18	1514	1517	1521	1524	1528	1531	1535	1538	1542	1545	0	1	1	1	2	2	2	3
·19	1549	1552	1556	1560	1563	1567	1570	1574	1578	1581	0	1	1	1	2	2	3	3
·20	1585	1589	1592	1596	1600	1603	1607	1611	1614	1618	0	1	1	1	2	2	3	3
·21	1622	1626	1629	1633	1637	1641	1644	1648	1652	1656	0	1	1	2	2	2	3	3
·22	1660	1663	1667	1671	1675	1679	1683	1687	1690	1694	0	1	1	2	2	2	3	3
·23	1698	1702	1706	1710	1714	1718	1722	1726	1730	1734	0	1	1	2	2	2	3	3
·24	1738	1742	1746	1750	1754	1758	1762	1766	1770	1774	0	1	1	2	2	2	3	3
·25	1778	1782	1786	1791	1795	1799	1803	1807	1811	1816	0	1	1	2	2	2	3	3
·26	1820	1824	1828	1832	1837	1841	1845	1849	1854	1858	0	1	1	2	2	3	3	3
·27	1862	1866	1871	1875	1879	1884	1888	1892	1897	1901	0	1	1	2	2	3	3	3
·28	1905	1910	1914	1919	1923	1928	1932	1936	1941	1945	0	1	1	2	2	3	3	4
·29	1950	1954	1959	1963	1968	1972	1977	1982	1986	1991	0	1	1	2	2	3	3	4
·30	1995	2000	2004	2009	2014	2018	2023	2028	2032	2037	0	1	1	2	2	3	3	4
·31	2042	2046	2051	2056	2061	2065	2070	2075	2080	2084	0	1	1	2	2	3	3	4
·32	2089	2094	2099	2104	2109	2113	2118	2123	2128	2133	0	1	1	2	2	3	3	4
·33	2138	2143	2148	2153	2158	2163	2168	2173	2178	2183	0	1	1	2	2	3	3	4
·34	2188	2193	2198	2203	2208	2213	2218	2223	2228	2234	1	1	2	2	3	3	4	4
·35	2239	2244	2249	2254	2259	2265	2270	2275	2280	2286	1	1	2	2	3	3	4	4
·36	2291	2296	2301	2307	2312	2317	2323	2328	2333	2339	1	1	2	2	3	3	4	4
·37	2344	2350	2355	2360	2366	2371	2377	2382	2388	2393	1	1	2	2	3	3	4	4
·38	2399	2404	2410	2415	2421	2427	2432	2438	2443	2449	1	1	2	2	3	3	4	4
·39	2455	2460	2466	2472	2477	2483	2489	2495	2500	2506	1	1	2	2	3	3	4	5
·40	2512	2518	2523	2529	2535	2541	2547	2553	2559	2564	1	1	2	2	3	4	4	5
·41	2570	2576	2582	2588	2594	2600	2606	2612	2618	2624	1	1	2	2	3	4	4	5
·42	2630	2636	2642	2649	2655	2661	2667	2673	2679	2685	1	1	2	2	3	4	4	5
·43	2692	2698	2704	2710	2716	2723	2729	2735	2742	2748	1	1	2	3	3	4	4	5
·44	2754	2761	2767	2773	2780	2786	2793	2799	2805	2812	1	1	2	3	3	4	4	5
·45	2818	2825	2831	2838	2844	2851	2858	2864	2871	2877	1	1	2	3	3	4	5	5
·46	2884	2891	2897	2904	2911	2917	2924	2931	2938	2944	1	1	2	3	3	4	5	5
·47	2951	2958	2965	2972	2979	2985	2992	2999	3006	3013	1	1	2	3	3	4	5	5
·48	3020	3027	3034	3041	3048	3055	3062	3069	3076	3083	1	1	2	3	4	4	5	6
·49	3090	3097	3105	3112	3119	3126	3133	3141	3148	3155	1	1	2	3	4	4	5	6
	0	1	2	3	4	5	6	7	8	9	1	2	3	4	5	6	7	8

	0	1	2	3	4	5	6	7	8	9	Differences 1	2	3	4	5	6	7	8	9
0	3162	3170	3177	3184	3192	3199	3206	3214	3221	3228	1	1	2	3	4	4	5	6	7
1	3236	3243	3251	3258	3266	3273	3281	3289	3296	3304	1	2	2	3	4	5	5	6	7
2	3311	3319	3327	3334	3342	3350	3357	3365	3373	3381	1	2	2	3	4	5	5	6	7
3	3388	3396	3404	3412	3420	3428	3436	3443	3451	3459	1	2	2	3	4	5	6	6	7
4	3467	3475	3483	3491	3499	3508	3516	3524	3532	3540	1	2	2	3	4	5	6	6	7
5	3548	3556	3565	3573	3581	3589	3597	3606	3614	3622	1	2	2	3	4	5	6	7	7
6	3631	3639	3648	3656	3664	3673	3681	3690	3698	3707	1	2	3	3	4	5	6	7	8
7	3715	3724	3733	3741	3750	3758	3767	3776	3784	3793	1	2	3	3	4	5	6	7	8
8	3802	3811	3819	3828	3837	3846	3855	3864	3873	3882	1	2	3	4	4	5	6	7	8
9	3890	3899	3908	3917	3926	3936	3945	3954	3963	3972	1	2	3	4	5	5	6	7	8
0	3981	3990	3999	4009	4018	4027	4036	4046	4055	4064	1	2	3	4	5	6	6	7	8
1	4074	4083	4093	4102	4111	4121	4130	4140	4150	4159	1	2	3	4	5	6	7	8	9
2	4169	4178	4188	4198	4207	4217	4227	4236	4246	4256	1	2	3	4	5	6	7	8	9
3	4266	4276	4285	4295	4305	4315	4325	4335	4345	4355	1	2	3	4	5	6	7	8	9
4	4365	4375	4385	4395	4406	4416	4426	4436	4446	4457	1	2	3	4	5	6	7	8	9
5	4467	4477	4487	4498	4508	4519	4529	4539	4550	4560	1	2	3	4	5	6	7	8	9
6	4571	4581	4592	4603	4613	4624	4634	4645	4656	4667	1	2	3	4	5	6	7	8	10
7	4677	4688	4699	4710	4721	4732	4742	4753	4764	4775	1	2	3	4	5	7	8	9	10
8	4786	4797	4808	4819	4831	4842	4853	4864	4875	4887	1	2	3	4	6	7	8	9	10
9	4898	4909	4920	4932	4943	4955	4966	4977	4989	5000	1	2	3	5	6	7	8	9	10
0	5012	5023	5035	5047	5058	5070	5082	5093	5105	5117	1	2	4	5	6	7	8	9	11
1	5129	5140	5152	5164	5176	5188	5200	5212	5224	5236	1	2	4	5	6	7	8	10	11
2	5248	5260	5272	5284	5297	5309	5321	5333	5346	5358	1	2	4	5	6	7	9	10	11
3	5370	5383	5395	5408	5420	5433	5445	5458	5470	5483	1	3	4	5	6	8	9	10	11
4	5495	5508	5521	5534	5546	5559	5572	5585	5598	5610	1	3	4	5	6	8	9	10	12
5	5623	5636	5649	5662	5675	5689	5702	5715	5728	5741	1	3	4	5	7	8	9	10	12
6	5754	5768	5781	5794	5808	5821	5834	5848	5861	5875	1	3	4	5	7	8	9	11	12
7	5888	5902	5916	5929	5943	5957	5970	5984	5998	6012	1	3	4	5	7	8	10	11	12
8	6026	6039	6053	6067	6081	6095	6109	6124	6138	6152	1	3	4	6	7	8	10	11	13
9	6166	6180	6194	6209	6223	6237	6252	6266	6281	6295	1	3	4	6	7	9	10	11	13
0	6310	6324	6339	6353	6368	6383	6397	6412	6427	6442	1	3	4	6	7	9	10	12	13
1	6457	6471	6486	6501	6516	6531	6546	6561	6577	6592	2	3	5	6	8	9	11	12	14
2	6607	6622	6637	6653	6668	6683	6699	6714	6730	6745	2	3	5	6	8	9	11	12	14
3	6761	6776	6792	6808	6823	6839	6855	6871	6887	6902	2	3	5	6	8	9	11	13	14
4	6918	6934	6950	6966	6982	6998	7015	7031	7047	7063	2	3	5	6	8	10	11	13	15
5	7079	7096	7112	7129	7145	7161	7178	7194	7211	7228	2	3	5	7	8	10	12	13	15
6	7244	7261	7278	7295	7311	7328	7345	7362	7379	7396	2	3	5	7	8	10	12	13	15
7	7413	7430	7447	7464	7482	7499	7516	7534	7551	7568	2	3	5	7	9	10	12	14	16
8	7586	7603	7621	7638	7656	7674	7691	7709	7727	7745	2	4	5	7	9	11	12	14	16
9	7762	7780	7798	7816	7834	7852	7870	7889	7907	7925	2	4	5	7	9	11	13	14	16
0	7943	7962	7980	7998	8017	8035	8054	8072	8091	8110	2	4	6	7	9	11	13	15	17
1	8128	8147	8166	8185	8204	8222	8241	8260	8279	8299	2	4	6	8	9	11	13	15	17
2	8318	8337	8356	8375	8395	8414	8433	8453	8472	8492	2	4	6	8	10	12	14	15	17
3	8511	8531	8551	8570	8590	8610	8630	8650	8670	8690	2	4	6	8	10	12	14	16	18
4	8710	8730	8750	8770	8790	8810	8831	8851	8872	8892	2	4	6	8	10	12	14	16	18
5	8913	8933	8954	8974	8995	9016	9036	9057	9078	9099	2	4	6	8	10	12	14	17	19
6	9120	9141	9162	9183	9204	9226	9247	9268	9290	9311	2	4	6	8	11	13	15	17	19
7	9333	9354	9376	9397	9419	9441	9462	9484	9506	9528	2	4	7	9	11	13	15	17	20
8	9550	9572	9594	9616	9638	9661	9683	9705	9727	9750	2	4	7	9	11	13	16	18	20
9	9772	9795	9817	9840	9863	9886	9908	9931	9954	9977	2	5	7	9	11	14	16	18	20
	0	1	2	3	4	5	6	7	8	9	1	2	3	4	5	6	7	8	9

	0	1	2	3	4	5	6	7	8	9	Differences							
											1	2	3	4	5	6	7	8
10	1000	1020	1040	1061	1082	1103	1124	1145	1166	1188	2	4	6	8	10	13	15	17
11	1210	1232	1254	1277	1300	1323	1346	1369	1392	1416	2	5	7	9	11	14	16	18
12	1440	1464	1488	1513	1538	1563	1588	1613	1638	1664	2	5	7	10	12	15	17	20
13	1690	1716	1742	1769	1796	1823	1850	1877	1904	1932	3	5	8	11	13	16	19	22
14	1960	1988	2016	2045	2074	2103	2132	2161	2190	2220	3	6	9	12	14	17	20	23
15	2250	2280	2310	2341	2372	2403	2434	2465	2496	2528	3	6	9	12	15	19	22	25
16	2560	2592	2624	2657	2690	2723	2756	2789	2822	2856	3	7	10	13	16	20	23	26
17	2890	2924	2958	2993	3028	3063	3098	3133	3168	3204	3	7	10	14	17	21	24	28
18	3240	3276	3312	3349	3386	3423	3460	3497	3534	3572	4	7	11	15	18	22	26	30
19	3610	3648	3686	3725	3764	3803	3842	3881	3920	3960	4	8	12	16	19	23	27	31
20	4000	4040	4080	4121	4162	4203	4244	4285	4326	4368	4	8	12	16	20	25	29	33
21	4410	4452	4494	4537	4580	4623	4666	4709	4752	4796	4	9	13	17	21	26	30	34
22	4840	4884	4928	4973	5018	5063	5108	5153	5198	5244	4	9	13	18	22	27	31	36
23	5290	5336	5382	5429	5476	5523	5570	5617	5664	5712	5	9	14	19	23	28	33	38
24	5760	5808	5856	5905	5954	6003	6052	6101	6150	6200	5	10	15	20	24	29	34	39
25	6250	6300	6350	6401	6452	6503	6554	6605	6656	6708	5	10	15	20	25	31	36	41
26	6760	6812	6864	6917	6970	7023	7076	7129	7182	7236	5	11	16	21	26	32	37	42
27	7290	7344	7398	7453	7508	7563	7618	7673	7728	7784	5	11	16	22	27	33	38	44
28	7840	7896	7952	8009	8066	8123	8180	8237	8294	8352	6	11	17	23	28	34	40	46
29	8410	8468	8526	8585	8644	8703	8762	8821	8880	8940	6	12	18	24	29	35	41	47
30	9000	9060	9120	9181	9242	9303	9364	9425	9486	9548	6	12	18	24	30	37	43	49
31	9610	9672	9734	9797	9860	9923	9986	—	—	—	6	13	19	25	31	38	44	50
31	—	—	—	—	—	—	—	1005	1011	1018	1	1	2	3	3	4	5	5
32	1024	1030	1037	1043	1050	1056	1063	1069	1076	1082	1	1	2	3	3	4	5	5
33	1089	1096	1102	1109	1116	1122	1129	1136	1142	1149	1	1	2	3	3	4	5	5
34	1156	1163	1170	1176	1183	1190	1197	1204	1211	1218	1	1	2	3	3	4	5	6
35	1225	1232	1239	1246	1253	1260	1267	1274	1282	1289	1	1	2	3	4	4	5	6
36	1296	1303	1310	1318	1325	1332	1340	1347	1354	1362	1	1	2	3	4	4	5	6
37	1369	1376	1384	1391	1399	1406	1414	1421	1429	1436	1	2	2	3	4	5	5	6
38	1444	1452	1459	1467	1475	1482	1490	1498	1505	1513	1	2	2	3	4	5	5	6
39	1521	1529	1537	1544	1552	1560	1568	1576	1584	1592	1	2	2	3	4	5	6	6
40	1600	1608	1616	1624	1632	1640	1648	1656	1665	1673	1	2	2	3	4	5	6	6
41	1681	1689	1697	1706	1714	1722	1731	1739	1747	1756	1	2	2	3	4	5	6	7
42	1764	1772	1781	1789	1798	1806	1815	1823	1832	1840	1	2	3	3	4	5	6	7
43	1849	1858	1866	1875	1884	1892	1901	1910	1918	1927	1	2	3	3	4	5	6	7
44	1936	1945	1954	1962	1971	1980	1989	1998	2007	2016	1	2	3	4	5	5	6	7
45	2025	2034	2043	2052	2061	2070	2079	2088	2098	2107	1	2	3	4	5	5	6	7
46	2116	2125	2134	2144	2153	2162	2172	2181	2190	2200	1	2	3	4	5	6	7	7
47	2209	2218	2228	2237	2247	2256	2266	2275	2285	2294	1	2	3	4	5	6	7	8
48	2304	2314	2323	2333	2343	2352	2362	2372	2381	2391	1	2	3	4	5	6	7	8
49	2401	2411	2421	2430	2440	2450	2460	2470	2480	2490	1	2	3	4	5	6	7	8
50	2500	2510	2520	2530	2540	2550	2560	2570	2581	2591	1	2	3	4	5	6	7	8
51	2601	2611	2621	2632	2642	2652	2663	2673	2683	2694	1	2	3	4	5	6	7	8
52	2704	2714	2725	2735	2746	2756	2767	2777	2788	2798	1	2	3	4	5	6	7	8
53	2809	2820	2830	2841	2852	2862	2873	2884	2894	2905	1	2	3	4	5	6	7	9
54	2916	2927	2938	2948	2959	2970	2981	2992	3003	3014	1	2	3	4	5	7	8	9

	0	1	2	3	4	5	6	7	8	9	Differences								
											1	2	3	4	5	6	7	8	9
5	3025	3036	3047	3058	3069	3080	3091	3102	3114	3125	1	2	3	4	6	7	8	9	10
6	3136	3147	3158	3170	3181	3192	3204	3215	3226	3238	1	2	3	5	6	7	8	9	10
7	3249	3260	3272	3283	3295	3306	3318	3329	3341	3352	1	2	3	5	6	7	8	9	10
8	3364	3376	3387	3399	3411	3422	3434	3446	3457	3469	1	2	4	5	6	7	8	9	11
9	3481	3493	3505	3516	3528	3540	3552	3564	3576	3588	1	2	4	5	6	7	8	10	11
0	3600	3612	3624	3636	3648	3660	3672	3684	3697	3709	1	2	4	5	6	7	8	10	11
1	3721	3733	3745	3758	3770	3782	3795	3807	3819	3832	1	2	4	5	6	7	9	10	11
2	3844	3856	3869	3881	3894	3906	3919	3931	3944	3956	1	2	4	5	6	7	9	10	11
3	3969	3982	3994	4007	4020	4032	4045	4058	4070	4083	1	3	4	5	6	8	9	10	11
4	4096	4109	4122	4134	4147	4160	4173	4186	4199	4212	1	3	4	5	6	8	9	10	12
5	4225	4238	4251	4264	4277	4290	4303	4316	4330	4343	1	3	4	5	7	8	9	10	12
6	4356	4369	4382	4396	4409	4422	4436	4449	4462	4476	1	3	4	5	7	8	9	11	12
7	4489	4502	4516	4529	4543	4556	4570	4583	4597	4610	1	3	4	5	7	8	9	11	12
8	4624	4638	4651	4665	4679	4692	4706	4720	4733	4747	1	3	4	5	7	8	10	11	12
9	4761	4775	4789	4802	4816	4830	4844	4858	4872	4886	1	3	4	6	7	8	10	11	13
0	4900	4914	4928	4942	4956	4970	4984	4998	5013	5027	1	3	4	6	7	8	10	11	13
1	5041	5055	5069	5084	5098	5112	5127	5141	5155	5170	1	3	4	6	7	9	10	11	13
2	5184	5198	5213	5227	5242	5256	5271	5285	5300	5314	1	3	4	6	7	9	10	11	13
3	5329	5344	5358	5373	5388	5402	5417	5432	5446	5461	1	3	4	6	7	9	10	12	13
4	5476	5491	5506	5520	5535	5550	5565	5580	5595	5610	1	3	4	6	7	9	10	12	13
5	5625	5640	5655	5670	5685	5700	5715	5730	5746	5761	2	3	5	6	8	9	11	12	14
6	5776	5791	5806	5822	5837	5852	5868	5883	5898	5914	2	3	5	6	8	9	11	12	14
7	5929	5944	5960	5975	5991	6006	6022	6037	6053	6068	2	3	5	6	8	9	11	12	14
8	6084	6100	6115	6131	6147	6162	6178	6194	6209	6225	2	3	5	6	8	9	11	13	14
9	6241	6257	6273	6288	6304	6320	6336	6352	6368	6384	2	3	5	6	8	10	11	13	14
0	6400	6416	6432	6448	6464	6480	6496	6512	6529	6545	2	3	5	6	8	10	11	13	14
1	6561	6577	6593	6610	6626	6642	6659	6675	6691	6708	2	3	5	7	8	10	11	13	15
2	6724	6740	6757	6773	6790	6806	6823	6839	6856	6872	2	3	5	7	8	10	12	13	15
3	6889	6906	6922	6939	6956	6972	6989	7006	7022	7039	2	3	5	7	8	10	12	13	15
4	7056	7073	7090	7106	7123	7140	7157	7174	7191	7208	2	3	5	7	8	10	12	14	15
5	7225	7242	7259	7276	7293	7310	7327	7344	7362	7379	2	3	5	7	9	10	12	14	15
6	7396	7413	7430	7448	7465	7482	7500	7517	7534	7552	2	3	5	7	9	10	12	14	16
7	7569	7586	7604	7621	7639	7656	7674	7691	7709	7726	2	3	5	7	9	10	12	14	16
8	7744	7762	7779	7797	7815	7832	7850	7868	7885	7903	2	4	5	7	9	11	12	14	16
9	7921	7939	7957	7974	7992	8010	8028	8046	8064	8082	2	4	5	7	9	11	13	14	16
0	8100	8118	8136	8154	8172	8190	8208	8226	8245	8263	2	4	5	7	9	11	13	14	16
1	8281	8299	8317	8336	8354	8372	8391	8409	8427	8446	2	4	5	7	9	11	13	15	16
2	8464	8482	8501	8519	8538	8556	8575	8593	8612	8630	2	4	6	7	9	11	13	15	17
3	8649	8668	8686	8705	8724	8742	8761	8780	8798	8817	2	4	6	7	9	11	13	15	17
4	8836	8855	8874	8892	8911	8930	8949	8968	8987	9006	2	4	6	8	9	11	13	15	17
5	9025	9044	9063	9082	9101	9120	9139	9158	9178	9197	2	4	6	8	10	11	13	15	17
6	9216	9235	9254	9274	9293	9312	9332	9351	9370	9390	2	4	6	8	10	12	14	15	17
7	9409	9428	9448	9467	9487	9506	9526	9545	9565	9584	2	4	6	8	10	12	14	16	18
8	9604	9624	9643	9663	9683	9702	9722	9742	9761	9781	2	4	6	8	10	12	14	16	18
9	9801	9821	9841	9860	9880	9900	9920	9940	9960	9980	2	4	6	8	10	12	14	16	18

	0	1	2	3	4	5	6	7	8	9	Differences 1	2	3	4	5	6	7	8
1·0	1·000	1·005	1·010	1·015	1·020	1·025	1·030	1·034	1·039	1·044	0	1	1	2	2	3	3	4
1·1	1·049	1·054	1·058	1·063	1·068	1·072	1·077	1·082	1·086	1·091	0	1	1	2	2	3	3	4
1·2	1·095	1·100	1·105	1·109	1·114	1·118	1·122	1·127	1·131	1·136	0	1	1	2	2	3	3	4
1·3	1·140	1·145	1·149	1·153	1·158	1·162	1·166	1·170	1·175	1·179	0	1	1	2	2	3	3	3
1·4	1·183	1·187	1·192	1·196	1·200	1·204	1·208	1·212	1·217	1·221	0	1	1	2	2	2	3	3
1·5	1·225	1·229	1·233	1·237	1·241	1·245	1·249	1·253	1·257	1·261	0	1	1	2	2	2	3	3
1·6	1·265	1·269	1·273	1·277	1·281	1·285	1·288	1·292	1·296	1·300	0	1	1	2	2	2	3	3
1·7	1·304	1·308	1·311	1·315	1·319	1·323	1·327	1·330	1·334	1·338	0	1	1	2	2	2	3	3
1·8	1·342	1·345	1·349	1·353	1·356	1·360	1·364	1·367	1·371	1·375	0	1	1	1	2	2	3	3
1·9	1·378	1·382	1·386	1·389	1·393	1·396	1·400	1·404	1·407	1·411	0	1	1	1	2	2	3	3
2·0	1·414	1·418	1·421	1·425	1·428	1·432	1·435	1·439	1·442	1·446	0	1	1	1	2	2	2	3
2·1	1·449	1·453	1·456	1·459	1·463	1·466	1·470	1·473	1·476	1·480	0	1	1	1	2	2	2	3
2·2	1·483	1·487	1·490	1·493	1·497	1·500	1·503	1·507	1·510	1·513	0	1	1	1	2	2	2	3
2·3	1·517	1·520	1·523	1·526	1·530	1·533	1·536	1·539	1·543	1·546	0	1	1	1	2	2	2	3
2·4	1·549	1·552	1·556	1·559	1·562	1·565	1·568	1·572	1·575	1·578	0	1	1	1	2	2	2	3
2·5	1·581	1·584	1·587	1·591	1·594	1·597	1·600	1·603	1·606	1·609	0	1	1	1	2	2	2	3
2·6	1·612	1·616	1·619	1·622	1·625	1·628	1·631	1·634	1·637	1·640	0	1	1	1	2	2	2	2
2·7	1·643	1·646	1·649	1·652	1·655	1·658	1·661	1·664	1·667	1·670	0	1	1	1	2	2	2	2
2·8	1·673	1·676	1·679	1·682	1·685	1·688	1·691	1·694	1·697	1·700	0	1	1	1	1	2	2	2
2·9	1·703	1·706	1·709	1·712	1·715	1·718	1·720	1·723	1·726	1·729	0	1	1	1	1	2	2	2
3·0	1·732	1·735	1·738	1·741	1·744	1·746	1·749	1·752	1·755	1·758	0	1	1	1	1	2	2	2
3·1	1·761	1·764	1·766	1·769	1·772	1·775	1·778	1·780	1·783	1·786	0	1	1	1	1	2	2	2
3·2	1·789	1·792	1·794	1·797	1·800	1·803	1·806	1·808	1·811	1·814	0	1	1	1	1	2	2	2
3·3	1·817	1·819	1·822	1·825	1·828	1·830	1·833	1·836	1·838	1·841	0	1	1	1	1	2	2	2
3·4	1·844	1·847	1·849	1·852	1·855	1·857	1·860	1·863	1·865	1·868	0	1	1	1	1	2	2	2
3·5	1·871	1·873	1·876	1·879	1·881	1·884	1·887	1·889	1·892	1·895	0	1	1	1	1	2	2	2
3·6	1·897	1·900	1·903	1·905	1·908	1·910	1·913	1·916	1·918	1·921	0	1	1	1	1	2	2	2
3·7	1·924	1·926	1·929	1·931	1·934	1·936	1·939	1·942	1·944	1·947	0	1	1	1	1	2	2	2
3·8	1·949	1·952	1·954	1·957	1·960	1·962	1·965	1·967	1·970	1·972	0	1	1	1	1	2	2	2
3·9	1·975	1·977	1·980	1·982	1·985	1·987	1·990	1·992	1·995	1·997	0	1	1	1	1	2	2	2
4·0	2·000	2·002	2·005	2·007	2·010	2·012	2·015	2·017	2·020	2·022	0	0	1	1	1	1	2	2
4·1	2·025	2·027	2·030	2·032	2·035	2·037	2·040	2·042	2·045	2·047	0	0	1	1	1	1	2	2
4·2	2·049	2·052	2·054	2·057	2·059	2·062	2·064	2·066	2·069	2·071	0	0	1	1	1	1	2	2
4·3	2·074	2·076	2·078	2·081	2·083	2·086	2·088	2·090	2·093	2·095	0	0	1	1	1	1	2	2
4·4	2·098	2·100	2·102	2·105	2·107	2·110	2·112	2·114	2·117	2·119	0	0	1	1	1	1	2	2
4·5	2·121	2·124	2·126	2·128	2·131	2·133	2·135	2·138	2·140	2·142	0	0	1	1	1	1	2	2
4·6	2·145	2·147	2·149	2·152	2·154	2·156	2·159	2·161	2·163	2·166	0	0	1	1	1	1	2	2
4·7	2·168	2·170	2·173	2·175	2·177	2·179	2·182	2·184	2·186	2·189	0	0	1	1	1	1	2	2
4·8	2·191	2·193	2·195	2·198	2·200	2·202	2·205	2·207	2·209	2·211	0	0	1	1	1	1	2	2
4·9	2·214	2·216	2·218	2·220	2·223	2·225	2·227	2·229	2·232	2·234	0	0	1	1	1	1	2	2
5·0	2·236	2·238	2·241	2·243	2·245	2·247	2·249	2·252	2·254	2·256	0	0	1	1	1	1	2	2
5·1	2·258	2·261	2·263	2·265	2·267	2·269	2·272	2·274	2·276	2·278	0	0	1	1	1	1	2	2
5·2	2·280	2·283	2·285	2·287	2·289	2·291	2·293	2·296	2·298	2·300	0	0	1	1	1	1	2	2
5·3	2·302	2·304	2·307	2·309	2·311	2·313	2·315	2·317	2·319	2·322	0	0	1	1	1	1	2	2
5·4	2·324	2·326	2·328	2·330	2·332	2·335	2·337	2·339	2·341	2·343	0	0	1	1	1	1	1	2

	0	1	2	3	4	5	6	7	8	9	Differences								
											1	2	3	4	5	6	7	8	9
·5	2·345	2·347	2·349	2·352	2·354	2·356	2·358	2·360	2·362	2·364	0	0	1	1	1	1	1	2	2
·6	2·366	2·369	2·371	2·373	2·375	2·377	2·379	2·381	2·383	2·385	0	0	1	1	1	1	1	2	2
·7	2·387	2·390	2·392	2·394	2·396	2·398	2·400	2·402	2·404	2·406	0	0	1	1	1	1	1	2	2
·8	2·408	2·410	2·412	2·415	2·417	2·419	2·421	2·423	2·425	2·427	0	0	1	1	1	1	1	2	2
·9	2·429	2·431	2·433	2·435	2·437	2·439	2·441	2·443	2·445	2·447	0	0	1	1	1	1	1	2	2
·0	2·449	2·452	2·454	2·456	2·458	2·460	2·462	2·464	2·466	2·468	0	0	1	1	1	1	1	2	2
·1	2·470	2·472	2·474	2·476	2·478	2·480	2·482	2·484	2·486	2·488	0	0	1	1	1	1	1	2	2
·2	2·490	2·492	2·494	2·496	2·498	2·500	2·502	2·504	2·506	2·508	0	0	1	1	1	1	1	2	2
·3	2·510	2·512	2·514	2·516	2·518	2·520	2·522	2·524	2·526	2·528	0	0	1	1	1	1	1	2	2
·4	2·530	2·532	2·534	2·536	2·538	2·540	2·542	2·544	2·546	2·548	0	0	1	1	1	1	1	2	2
·5	2·550	2·551	2·553	2·555	2·557	2·559	2·561	2·563	2·565	2·567	0	0	1	1	1	1	1	2	2
·6	2·569	2·571	2·573	2·575	2·577	2·579	2·581	2·583	2·585	2·587	0	0	1	1	1	1	1	2	2
·7	2·588	2·590	2·592	2·594	2·596	2·598	2·600	2·602	2·604	2·606	0	0	1	1	1	1	1	2	2
·8	2·608	2·610	2·612	2·613	2·615	2·617	2·619	2·621	2·623	2·625	0	0	1	1	1	1	1	2	2
·9	2·627	2·629	2·631	2·632	2·634	2·636	2·638	2·640	2·642	2·644	0	0	1	1	1	1	1	2	2
·0	2·646	2·648	2·650	2·651	2·653	2·655	2·657	2·659	2·661	2·663	0	0	1	1	1	1	1	2	2
·1	2·665	2·666	2·668	2·670	2·672	2·674	2·676	2·678	2·680	2·681	0	0	1	1	1	1	1	1	2
·2	2·683	2·685	2·687	2·689	2·691	2·693	2·694	2·696	2·698	2·700	0	0	1	1	1	1	1	1	2
·3	2·702	2·704	2·706	2·707	2·709	2·711	2·713	2·715	2·717	2·718	0	0	1	1	1	1	1	1	2
·4	2·720	2·722	2·724	2·726	2·728	2·729	2·731	2·733	2·735	2·737	0	0	1	1	1	1	1	1	2
·5	2·739	2·740	2·742	2·744	2·746	2·748	2·750	2·751	2·753	2·755	0	0	1	1	1	1	1	1	2
·6	2·757	2·759	2·760	2·762	2·764	2·766	2·768	2·769	2·771	2·773	0	0	1	1	1	1	1	1	2
·7	2·775	2·777	2·778	2·780	2·782	2·784	2·786	2·787	2·789	2·791	0	0	1	1	1	1	1	1	2
·8	2·793	2·795	2·796	2·798	2·800	2·802	2·804	2·805	2·807	2·809	0	0	1	1	1	1	1	1	2
·9	2·811	2·812	2·814	2·816	2·818	2·820	2·821	2·823	2·825	2·827	0	0	1	1	1	1	1	1	2
·0	2·828	2·830	2·832	2·834	2·835	2·837	2·839	2·841	2·843	2·844	0	0	1	1	1	1	1	1	2
·1	2·846	2·848	2·850	2·851	2·853	2·855	2·857	2·858	2·860	2·862	0	0	1	1	1	1	1	1	2
·2	2·864	2·865	2·867	2·869	2·871	2·872	2·874	2·876	2·877	2·879	0	0	1	1	1	1	1	1	2
·3	2·881	2·883	2·884	2·886	2·888	2·890	2·891	2·893	2·895	2·897	0	0	1	1	1	1	1	1	2
·4	2·898	2·900	2·902	2·903	2·905	2·907	2·909	2·910	2·912	2·914	0	0	1	1	1	1	1	1	2
·5	2·915	2·917	2·919	2·921	2·922	2·924	2·926	2·927	2·929	2·931	0	0	1	1	1	1	1	1	2
·6	2·933	2·934	2·936	2·938	2·939	2·941	2·943	2·944	2·946	2·948	0	0	1	1	1	1	1	1	2
·7	2·950	2·951	2·953	2·955	2·956	2·958	2·960	2·961	2·963	2·965	0	0	1	1	1	1	1	1	2
·8	2·966	2·968	2·970	2·972	2·973	2·975	2·977	2·978	2·980	2·982	0	0	1	1	1	1	1	1	2
·9	2·983	2·985	2·987	2·988	2·990	2·992	2·993	2·995	2·997	2·998	0	0	1	1	1	1	1	1	2
·0	3·000	3·002	3·003	3·005	3·007	3·008	3·010	3·012	3·013	3·015	0	0	0	1	1	1	1	1	1
·1	3·017	3·018	3·020	3·022	3·023	3·025	3·027	3·028	3·030	3·032	0	0	0	1	1	1	1	1	1
·2	3·033	3·035	3·036	3·038	3·040	3·041	3·043	3·045	3·046	3·048	0	0	0	1	1	1	1	1	1
·3	3·050	3·051	3·053	3·055	3·056	3·058	3·059	3·061	3·063	3·064	0	0	0	1	1	1	1	1	1
·4	3·066	3·068	3·069	3·071	3·072	3·074	3·076	3·077	3·079	3·081	0	0	0	1	1	1	1	1	1
·5	3·082	3·084	3·085	3·087	3·089	3·090	3·092	3·094	3·095	3·097	0	0	0	1	1	1	1	1	1
·6	3·098	3·100	3·102	3·103	3·105	3·106	3·108	3·110	3·111	3·113	0	0	0	1	1	1	1	1	1
·7	3·114	3·116	3·118	3·119	3·121	3·122	3·124	3·126	3·127	3·129	0	0	0	1	1	1	1	1	1
·8	3·130	3·132	3·134	3·135	3·137	3·138	3·140	3·142	3·143	3·145	0	0	0	1	1	1	1	1	1
·9	3·146	3·148	3·150	3·151	3·153	3·154	3·156	3·158	3·159	3·161	0	0	0	1	1	1	1	1	1

	0	1	2	3	4	5	6	7	8	9	Differences 1	2	3	4	5	6	7	8	9
10	3·162	3·178	3·194	3·209	3·225	3·240	3·256	3·271	3·286	3·302	2	3	5	6	8	9	11	12	[illegible]
11	3·317	3·332	3·347	3·362	3·376	3·391	3·406	3·421	3·435	3·450	1	3	4	6	7	9	10	12	[illegible]
12	3·464	3·479	3·493	3·507	3·521	3·536	3·550	3·564	3·578	3·592	1	3	4	6	7	8	10	11	[illegible]
13	3·606	3·619	3·633	3·647	3·661	3·674	3·688	3·701	3·715	3·728	1	3	4	5	7	8	10	11	[illegible]
14	3·742	3·755	3·768	3·782	3·795	3·808	3·821	3·834	3·847	3·860	1	3	4	5	7	8	9	11	[illegible]
15	3·873	3·886	3·899	3·912	3·924	3·937	3·950	3·962	3·975	3·987	1	3	4	5	6	8	9	10	[illegible]
16	4·000	4·012	4·025	4·037	4·050	4·062	4·074	4·087	4·099	4·111	1	2	4	5	6	7	9	10	[illegible]
17	4·123	4·135	4·147	4·159	4·171	4·183	4·195	4·207	4·219	4·231	1	2	4	5	6	7	8	10	[illegible]
18	4·243	4·254	4·266	4·278	4·290	4·301	4·313	4·324	4·336	4·347	1	2	3	5	6	7	8	9	[illegible]
19	4·359	4·370	4·382	4·393	4·405	4·416	4·427	4·438	4·450	4·461	1	2	3	5	6	7	8	9	[illegible]
20	4·472	4·483	4·494	4·506	4·517	4·528	4·539	4·550	4·561	4·572	1	2	3	4	6	7	8	9	[illegible]
21	4·583	4·593	4·604	4·615	4·626	4·637	4·648	4·658	4·669	4·680	1	2	3	4	5	6	8	9	[illegible]
22	4·690	4·701	4·712	4·722	4·733	4·743	4·754	4·764	4·775	4·785	1	2	3	4	5	6	7	8	[illegible]
23	4·796	4·806	4·817	4·827	4·837	4·848	4·858	4·868	4·879	4·889	1	2	3	4	5	6	7	8	[illegible]
24	4·899	4·909	4·919	4·930	4·940	4·950	4·960	4·970	4·980	4·990	1	2	3	4	5	6	7	8	[illegible]
25	5·000	5·010	5·020	5·030	5·040	5·050	5·060	5·070	5·079	5·089	1	2	3	4	5	6	7	8	[illegible]
26	5·099	5·109	5·119	5·128	5·138	5·148	5·158	5·167	5·177	5·187	1	2	3	4	5	6	7	8	[illegible]
27	5·196	5·206	5·215	5·225	5·235	5·244	5·254	5·263	5·273	5·282	1	2	3	4	5	6	7	8	[illegible]
28	5·292	5·301	5·310	5·320	5·329	5·339	5·348	5·357	5·367	5·376	1	2	3	4	5	6	7	7	[illegible]
29	5·385	5·394	5·404	5·413	5·422	5·431	5·441	5·450	5·459	5·468	1	2	3	4	5	5	6	7	[illegible]
30	5·477	5·486	5·495	5·505	5·514	5·523	5·532	5·541	5·550	5·559	1	2	3	4	4	5	6	7	[illegible]
31	5·568	5·577	5·586	5·595	5·604	5·612	5·621	5·630	5·639	5·648	1	2	3	3	4	5	6	7	[illegible]
32	5·657	5·666	5·675	5·683	5·692	5·701	5·710	5·718	5·727	5·736	1	2	3	3	4	5	6	7	[illegible]
33	5·745	5·753	5·762	5·771	5·779	5·788	5·797	5·805	5·814	5·822	1	2	3	3	4	5	6	7	[illegible]
34	5·831	5·840	5·848	5·857	5·865	5·874	5·882	5·891	5·899	5·908	1	2	3	3	4	5	6	7	[illegible]
35	5·916	5·925	5·933	5·941	5·950	5·958	5·967	5·975	5·983	5·992	1	2	2	3	4	5	6	7	[illegible]
36	6·000	6·008	6·017	6·025	6·033	6·042	6·050	6·058	6·066	6·075	1	2	2	3	4	5	6	7	[illegible]
37	6·083	6·091	6·099	6·107	6·116	6·124	6·132	6·140	6·148	6·156	1	2	2	3	4	5	6	7	[illegible]
38	6·164	6·173	6·181	6·189	6·197	6·205	6·213	6·221	6·229	6·237	1	2	2	3	4	5	6	6	[illegible]
39	6·245	6·253	6·261	6·269	6·277	6·285	6·293	6·301	6·309	6·317	1	2	2	3	4	5	6	6	[illegible]
40	6·325	6·332	6·340	6·348	6·356	6·364	6·372	6·380	6·387	6·395	1	2	2	3	4	5	6	6	[illegible]
41	6·403	6·411	6·419	6·427	6·434	6·442	6·450	6·458	6·465	6·473	1	2	2	3	4	5	5	6	[illegible]
42	6·481	6·488	6·496	6·504	6·512	6·519	6·527	6·535	6·542	6·550	1	2	2	3	4	5	5	6	[illegible]
43	6·557	6·565	6·573	6·580	6·588	6·595	6·603	6·611	6·618	6·626	1	2	2	3	4	5	5	6	[illegible]
44	6·633	6·641	6·648	6·656	6·663	6·671	6·678	6·686	6·693	6·701	1	1	2	3	4	4	5	6	[illegible]
45	6·708	6·716	6·723	6·731	6·738	6·745	6·753	6·760	6·768	6·775	1	1	2	3	4	4	5	6	[illegible]
46	6·782	6·790	6·797	6·804	6·812	6·819	6·826	6·834	6·841	6·848	1	1	2	3	4	4	5	6	[illegible]
47	6·856	6·863	6·870	6·877	6·885	6·892	6·899	6·907	6·914	6·921	1	1	2	3	4	4	5	6	[illegible]
48	6·928	6·935	6·943	6·950	6·957	6·964	6·971	6·979	6·986	6·993	1	1	2	3	4	4	5	6	[illegible]
49	7·000	7·007	7·014	7·021	7·029	7·036	7·043	7·050	7·057	7·064	1	1	2	3	4	4	5	6	[illegible]
50	7·071	7·078	7·085	7·092	7·099	7·106	7·113	7·120	7·127	7·134	1	1	2	3	4	4	5	6	[illegible]
51	7·141	7·148	7·155	7·162	7·169	7·176	7·183	7·190	7·197	7·204	1	1	2	3	4	4	5	6	[illegible]
52	7·211	7·218	7·225	7·232	7·239	7·246	7·253	7·259	7·266	7·273	1	1	2	3	3	4	5	6	[illegible]
53	7·280	7·287	7·294	7·301	7·308	7·314	7·321	7·328	7·335	7·342	1	1	2	3	3	4	5	5	[illegible]
54	7·348	7·355	7·362	7·369	7·376	7·382	7·389	7·396	7·403	7·409	1	1	2	3	3	4	5	5	[illegible]

	0	1	2	3	4	5	6	7	8	9	Differences 1	2	3	4	5	6	7	8	9
55	7·416	7·423	7·430	7·436	7·443	7·450	7·457	7·463	7·470	7·477	1	1	2	3	3	4	5	5	6
56	7·483	7·490	7·497	7·503	7·510	7·517	7·523	7·530	7·537	7·543	1	1	2	3	3	4	5	5	6
57	7·550	7·556	7·563	7·570	7·576	7·583	7·589	7·596	7·603	7·609	1	1	2	3	3	4	5	5	6
58	7·616	7·622	7·629	7·635	7·642	7·649	7·655	7·662	7·668	7·675	1	1	2	3	3	4	5	5	6
59	7·681	7·688	7·694	7·701	7·707	7·714	7·720	7·727	7·733	7·740	1	1	2	3	3	4	5	5	6
60	7·746	7·752	7·759	7·765	7·772	7·778	7·785	7·791	7·797	7·804	1	1	2	3	3	4	4	5	6
61	7·810	7·817	7·823	7·829	7·836	7·842	7·849	7·855	7·861	7·868	1	1	2	3	3	4	4	5	6
62	7·874	7·880	7·887	7·893	7·899	7·906	7·912	7·918	7·925	7·931	1	1	2	3	3	4	4	5	6
63	7·937	7·944	7·950	7·956	7·962	7·969	7·975	7·981	7·987	7·994	1	1	2	3	3	4	4	5	6
64	8·000	8·006	8·012	8·019	8·025	8·031	8·037	8·044	8·050	8·056	1	1	2	2	3	4	4	5	6
65	8·062	8·068	8·075	8·081	8·087	8·093	8·099	8·106	8·112	8·118	1	1	2	2	3	4	4	5	6
66	8·124	8·130	8·136	8·142	8·149	8·155	8·161	8·167	8·173	8·179	1	1	2	2	3	4	4	5	5
67	8·185	8·191	8·198	8·204	8·210	8·216	8·222	8·228	8·234	8·240	1	1	2	2	3	4	4	5	5
68	8·246	8·252	8·258	8·264	8·270	8·276	8·283	8·289	8·295	8·301	1	1	2	2	3	4	4	5	5
69	8·307	8·313	8·319	8·325	8·331	8·337	8·343	8·349	8·355	8·361	1	1	2	2	3	4	4	5	5
70	8·367	8·373	8·379	8·385	8·390	8·396	8·402	8·408	8·414	8·420	1	1	2	2	3	4	4	5	5
71	8·426	8·432	8·438	8·444	8·450	8·456	8·462	8·468	8·473	8·479	1	1	2	2	3	4	4	5	5
72	8·485	8·491	8·497	8·503	8·509	8·515	8·521	8·526	8·532	8·538	1	1	2	2	3	3	4	5	5
73	8·544	8·550	8·556	8·562	8·567	8·573	8·579	8·585	8·591	8·597	1	1	2	2	3	3	4	5	5
74	8·602	8·608	8·614	8·620	8·626	8·631	8·637	8·643	8·649	8·654	1	1	2	2	3	3	4	5	5
75	8·660	8·666	8·672	8·678	8·683	8·689	8·695	8·701	8·706	8·712	1	1	2	2	3	3	4	5	5
76	8·718	8·724	8·729	8·735	8·741	8·746	8·752	8·758	8·764	8·769	1	1	2	2	3	3	4	5	5
77	8·775	8·781	8·786	8·792	8·798	8·803	8·809	8·815	8·820	8·826	1	1	2	2	3	3	4	4	5
78	8·832	8·837	8·843	8·849	8·854	8·860	8·866	8·871	8·877	8·883	1	1	2	2	3	3	4	4	5
79	8·888	8·894	8·899	8·905	8·911	8·916	8·922	8·927	8·933	8·939	1	1	2	2	3	3	4	4	5
80	8·944	8·950	8·955	8·961	8·967	8·972	8·978	8·983	8·989	8·994	1	1	2	2	3	3	4	4	5
81	9·000	9·006	9·011	9·017	9·022	9·028	9·033	9·039	9·044	9·050	1	1	2	2	3	3	4	4	5
82	9·055	9·061	9·066	9·072	9·077	9·083	9·088	9·094	9·099	9·105	1	1	2	2	3	3	4	4	5
83	9·110	9·116	9·121	9·127	9·132	9·138	9·143	9·149	9·154	9·160	1	1	2	2	3	3	4	4	5
84	9·165	9·171	9·176	9·182	9·187	9·192	9·198	9·203	9·209	9·214	1	1	2	2	3	3	4	4	5
85	9·220	9·225	9·230	9·236	9·241	9·247	9·252	9·257	9·263	9·268	1	1	2	2	3	3	4	4	5
86	9·274	9·279	9·284	9·290	9·295	9·301	9·306	9·311	9·317	9·322	1	1	2	2	3	3	4	4	5
87	9·327	9·333	9·338	9·343	9·349	9·354	9·359	9·365	9·370	9·375	1	1	2	2	3	3	4	4	5
88	9·381	9·386	9·391	9·397	9·402	9·407	9·413	9·418	9·423	9·429	1	1	2	2	3	3	4	4	5
89	9·434	9·439	9·445	9·450	9·455	9·460	9·466	9·471	9·476	9·482	1	1	2	2	3	3	4	4	5
90	9·487	9·492	9·497	9·503	9·508	9·513	9·518	9·524	9·529	9·534	1	1	2	2	3	3	4	4	5
91	9·539	9·545	9·550	9·555	9·560	9·566	9·571	9·576	9·581	9·586	1	1	2	2	3	3	4	4	5
92	9·592	9·597	9·602	9·607	9·612	9·618	9·623	9·628	9·633	9·638	1	1	2	2	3	3	4	4	5
93	9·644	9·649	9·654	9·659	9·664	9·670	9·675	9·680	9·685	9·690	1	1	2	2	3	3	4	4	5
94	9·695	9·701	9·706	9·711	9·716	9·721	9·726	9·731	9·737	9·742	1	1	2	2	3	3	4	4	5
95	9·747	9·752	9·757	9·762	9·767	9·772	9·778	9·783	9·788	9·793	1	1	2	2	3	3	4	4	5
96	9·798	9·803	9·808	9·813	9·818	9·823	9·829	9·834	9·839	9·844	1	1	2	2	3	3	4	4	5
97	9·849	9·854	9·859	9·864	9·869	9·874	9·879	9·884	9·889	9·894	1	1	2	2	3	3	4	4	5
98	9·899	9·905	9·910	9·915	9·920	9·925	9·930	9·935	9·940	9·945	0	1	1	2	2	3	3	4	4
99	9·950	9·955	9·960	9·965	9·970	9·975	9·980	9·985	9·990	9·995	0	1	1	2	2	3	3	4	4